The

PRIMAL
FEAST

Also by Susan Allport

A Natural History of Parenting:
A Naturalist Looks at Parenting in the
Animal World and Ours

Sermons in Stone:
The Stone Walls of New England
and New York

Explorers of the Black Box:
The Search for the Cellular Basis
of Memory

The
PRIMAL
FEAST,

Food, Sex,

Foraging,

and Love

SUSAN ALLPORT

Harmony Books ⬥ New York

Published by Harmony Books, 201 East 50th Street, New York,
New York 10022. Member of the Crown Publishing Group.

Random House, Inc. New York, Toronto, London, Sydney, Auckland
www.randomhouse.com

HARMONY BOOKS is a registered trademark and the Harmony Books
colophon is a trademark of Random House, Inc.

Printed in the United States of America

Design by Barbara Sturman

Library of Congress Cataloging-in-Publication Data

Allport, Susan
    The primal feast : food, sex, foraging, and love / Susan Allport.—
1st ed.
    Includes bibliographic references.
    1. Food.   2. Food Habits.   I. Title
TX353 .A56 1999
641.3—dc21   99-046244

ISBN 0-609-60149-0

10 9 8 7 6 5 4 3 2 1

First Edition

FOR DAVID,
who always asks good questions,

AND FOR LIBERTY AND CECIL,
who began searching for wild foods
at a very young age

ACKNOWLEDGMENTS

I'd like to thank the many anthropologists, archaeologists, biologists, nutritionists, and foragers who contributed to this book and especially Tony Gaston and Ken Glander for the glimpses that they gave me of life on a remote island in the Arctic and a tropical forest in Costa Rica. Ted Sumner showed me where to find chanterelles in the woods in upstate New York; Wildman Steve Brill took me foraging in Prospect Park in Brooklyn and enabled me to see that the woods of Westchester had become a deciduous desert; and Ed Olsen gave me the use of his deer blind.

I am also grateful to Tony Gaston and Ken Glander, as well as to Wenda Trevathan, Barbara Rolls, Ellen Dierenfeld, Clark Larsen, Stevie Pierson, Suzanne Ironbiter, Mary Ann Tirone Smith, Ginger Barber, and my father Sandy Allport, for reading the book in manuscript form and offering their many invaluable suggestions; and to Debby Benz at the Katonah Village Library for supplying me with more articles and books than any single patron has the right to ask for. Shaye Areheart sensed that there was a book in my musings about food, and my family ate almost every one of my foraged dinners.

CONTENTS

If a rich man, when you will;
if a poor man, when you can.

DIOGENES LAERTIUS,
when asked the proper time to eat,
third century A.D.

One

AN EARLY-MORNING AROUSAL

I t is 5:45 in the morning, and I have been awakened, as I have the past two mornings, by the sounds of a squirrel in the gutter near my window. It is searching for dogwood berries, I think, because a very old dogwood hangs over the house and because I have seen squirrels during the day rummaging through this and other gutters and eating those bright-red, lozenge-shaped fruits. Their mouths are stained red. Their teeth move as rapidly as the needle on a sewing machine.

Isn't it too early for you? I want to ask this squirrel. No light at all is coming through the window at 5:45 on this day in mid-November. But I know the early squirrel gets the dogwood berry. That squirrels spend most of their time eating and searching for food is strange only to people who have their meals down to a regular three a day.

Anyway, I'm awake now, and it's still early enough to get up and do something I've been promising myself I'd do for several months. I dress quickly, putting on wool socks, long underwear, and the warm fleece pants that I wear ice skating. Then I head outside and down into the woods behind my house. The crescent moon is bright in the sky. The leafless trees stand bold and black against the first pale, pearly light.

I walk slowly so that I won't trip over any rocks or branches, but in fact I know this way well. It is the same path I take to the spring where I pick watercress or to the compost pile where my husband and I bring our wagonloads of clippings and prunings. Neither of those places is my destination, though, and I pass them both and begin to follow a stone wall that marks our property line down to the point where it intersects with another wall. There I see the tree that I intend to climb, a tulip tree about seventy-five feet high. It is split at its base into three tall straight trunks, and on one of the trunks, about a third of the way up, is a deer blind, a small metal seat placed there by a local carpenter who has our permission to hunt in these woods.

The carpenter also gave me permission to climb up to his isolated perch, though if he had any idea how much I like to have both feet on the ground and how little I like being up in something that sways when the wind blows, as this tree does, he might have done me the favor of saying no. I have climbed the tree twice now, during the day, and I know that the rungs screwed into the trunk are positioned so that the climb up is fairly easy. But still, one has, at the top, to hoist oneself up and over the wire supports of the blind.

And one has, of course, to come down. I would be enjoying myself a lot more if I didn't know that. The first time I tried it, I started off with my hands and feet in the wrong position and wound up hanging from one of the trunks until I mustered the nerve and the strength to pull myself back up to the top and start over. My heart was still pounding when I got home, and my muscles were sore for days.

This morning it is so dark that I can barely make out the rungs. When I find them, they are strikingly cold as only metal can be. I climb quickly and am soon settled on the grid seat, with my gloves on and my binoculars ready, waiting. I check the time. It is a few minutes after six, and I watch as the marsh on my left, and then the apple trees surrounding the spring where the watercress grows, come into focus. What changes this land has seen, I think from this high vantage point. What changes this land will continue to see. The stone walls, rock-lined spring, and apple trees remind me that just one hundred years ago these woods were farmland. The bare understory of the woods, the fact that white-tailed deer have eaten all the young saplings with which a forest regenerates itself, makes me wonder what these woods will look like in another hundred years. "No one knows," a Cornell wildlife specialist who has been studying the question once told me, "because we can't make any predictions from the past." He tells his students, though, that whatever the species composition is in the forests that their children and their grandchildren will walk through, it will be determined primarily by deer.

I look down at the forest floor, and the only low-growing plants I can see are white snakeroot with its flower heads, now fuzzy and brown, and winged euonymous, its small pale pink leaves hanging from its branches like the crystal drops on a chandelier. I don't know why the deer pass up euonymous, or corkbush as it is also called, but I do know that white snakeroot contains the powerful and deadly poison tremetol, a compound similar to rotenone, the active ingredient in many pesticides. Deer, like other herbivores, have a great ability to detoxify compounds that would fell other animals, but even they are no match for tremetol.

So, I think, we may not know what the forest will look like in one hundred years, but we do probably know what it will taste like. Whatever trees and plants are allowed to survive and reproduce will have leaves and bark deadly enough to put off deer. A density of just eighteen to twenty deer per square mile starts to significantly change the forest understory. In these parts of New York, as in much of the east coast, deer populations are five to ten times that. It would be wonderful to think of that much wildlife just forty miles from New York City were it not for the long-term effects on the land and on all the other animals that live in these woods and are dependent on a healthy mix of plants for shelter, nesting materials, and food.

I fully expected to see some of these deer this morning, some of the many scores of deer that I pass every day on my walks or while driving in my car. In fact, I came out this early just to get an appreciation of what a population density of some one hundred deer per square mile looks like during the morning hours, the deer rush hours. I expected to be able to muse on the fact that only here in the United States, only now in the late twentieth century, are food supplies so abundant and available that we don't have to bother with these convenient packets of protein and fat. And to point out that only because we are so removed from food production can we afford to let this one animal eliminate all the edible plants in our area, all the plants that we would need to survive if our supermarkets suddenly closed.

But my morning didn't turn out the way I thought it would. I didn't see any deer. I think I heard one warning snort of the kind that a deer makes to warn other deer that something is amiss, but

I didn't actually see a single white tail, a single flashy new set of antlers. Perhaps it was the bright blue socks that I had grabbed in the dark or my sneakers, which had never looked so white as they did dangling down from that seat. Or perhaps it was that the deer had already eaten everything there was to eat in this part of the woods and were on my lawn, sampling the rose of Sharon that I planted last summer and had forgotten to fence.

I don't know. But what I did come to appreciate that morning, as I sat there on my high seat, trying to tuck my feet underneath me so that my shoes and socks would not be so conspicuous, was just how difficult it would be to actually bring dinner home this way. I understand very well the skills that it takes to grow food or prepare food, but somehow I never really appreciated the skills that it takes to hunt food. I don't know what I was thinking: that hunters just set themselves up in trees, then pick off their quarry as the deer file by. But as I sat there growing a little hungry, cold, and tired of waiting, thinking of how much I'd rather be poking around in the spring for watercress, I was getting a new perspective on the traditionally male half of the human food-getting equation.

I found myself remembering some of the more remarkable hunting techniques I had heard about in the past year or so: hunters waiting for days at a spring or a fruiting tree; hunters in Kenya and Tanzania draping themselves in the red blankets of Masai herdsmen so they can move in close to zebra and gazelles without those animals suspecting their true intentions; Inuit hunters outwitting seals at their breathing holes with a method that resembles nothing so much as a cleverly orchestrated shell game. In it, two Inuit walk toward a breathing hole together and with carefully synchronized steps. Then, as they reach the hole, one stops and the other keeps on walking. The seal comes to think, naturally and mistakenly, that it is safe to poke its nose up for air.

Nature doesn't give up its truths easily, a scientist once told me after spending five thousand hours in the field collecting data on the habits of a particular monkey. Or its foods, I thought, as I finally gave up waiting and began climbing down the tree.

Or its foods.

Two

THE DINNER PARTY

These thoughts about food and the quest for food began one night a few years ago, over a wine-dark lamb stew and a question posed by my husband. We were having friends to dinner, and at some point during the evening, he asked us all what role we would have played in a much simpler, subsistence society. In truth, no one in such a society would have had the option of spending their days in any other way than in the near-constant quest for food. But my husband got across the idea that what he wanted from us was what we thought our deepest talents were, what we could contribute in a subsistence society and be happy in the doing.

My husband was sure how he would have been occupied ten or twenty thousand years ago. He would have been a toolmaker and a tinkerer. He would have been busy inventing new ways to build shelters, haul water, pick berries, and scrape the meat off carcasses. There were no declared hunters in our group that night;

our friends saw themselves as storytellers, shamans, or medicine men. But I immediately knew that I would have been a forager—a gatherer of wild foods.

I answered "scavenger" at dinner, but I didn't mean scavenger in the sense of one that feeds on dead animals. Rather, to scavenge or search for usable goods—in this case, food—at no cost. That's how I spent my summers when I was much younger and working at odd, poorly paid jobs on Cape Cod. During a bottle-washing stint at the Marine Biological Laboratories in Woods Hole, I lived in a tent and ate the giant clams called quahogs that a friend and I used to dive for, then stuff with various store-bought and wild ingredients. My favorite at the time was quahogs with black beans and wild onions, though I now suspect it was something that only a forager could love. Other summers, with Euell Gibbon's *Stalking the Wild Asparagus* as my guide, I fed my friends and myself on chowders and stews made from mussels and oysters I had collected and on pies filled with hand-picked huckleberries. By then I might have been a better cook (or at least I knew to add lots of butter and cream to my concoctions), and one of those summer chowders is still remembered longingly by a certain Epicurean in my family.

Still, I had never thought of myself as either a scavenger or a forager before that night. I was a writer, a wife, and a mother, and foraging was just something I do every now and then to put food—food that is a little different, a little more interesting—on the table. So I was surprised at how quickly the answer to my husband's question popped into my mind—and how sure I was that it was right. It made so much sense that I suddenly found myself counting the ways in which I knew it was true.

I know that I am a forager because I can't start work in the morning until I have settled the question of what I'll be cooking for dinner.

I know that I am a forager because although I can't remember where I parked my car in a parking lot, I can always remember the exact spot where I found a stand of ostrich fern, a patch of lemony curly dock.

I know that I am a forager because, during the summer while other people are off playing golf or tennis, I spend my time searching for mushrooms. There are others like me, I also know, because I've spotted them in the woods with their baskets and pails. I've read their articles on morelists and fine foraging in magazines. I've even seen a cartoon about them in *The New Yorker.* In it, three people are outside on a porch with rain pouring down around them. A man and a woman sit despondently with golf club and tennis racket in hand, as a very cheerful woman stands in the doorway. "Well, it *has* been a great summer for chanterelles," she says.

I know that I am a forager because the books I like tend to be about food and the effect that food can have on a person's destiny. I'm thinking, of course, of *Growth of the Soil, The Good Earth,* and the chapter in *Far From the Madding Crowd* in which Gabriel Oak's sheepdog eats a dead lamb and then uses his new-found vigor and energy to drive all of Oak's sheep over the cliff. Some of my favorite movies also have to do with food: *Eat, Drink, Man, Woman, La Grande Bouffe, Big Night,* and *Babette's Feast,* in which Babette, a French chef and refugee, teaches the people of a small village in Denmark that food is more than just nourishment. It is pleasure, forgiveness, gratitude, and love. And I can't forget *Ermo,* a film out of Communist China, in which a woman who makes twisted noodles winds up selling her body, blood, and soul in order to buy her son a television set. One of my favorite pieces of sculpture, carved out of quartz by an Inuit artist, is also about food. It is called *Four Hungry Bears Dreaming of a Whale,* and I saw it on Baffin Island when I was on my way to an island in

Hudson Bay to observe a colony of thick-billed murres. Four white bears are walking upright and in a straight line. A white whale seems to float above their heads.

I know that I am a forager because one of my favorite things about my house—a house with many charms—is that it has a small spring in the woods that is carpeted with fresh peppery watercress in the spring and the fall.

I know that I am a forager because my favorite color is green.

I know that I am a forager because of the profession I've chosen. For how better to describe science writing than foraging for discoveries in different fields, scavenging for the ideas and research of others?

Most of all, however, I know that I'm a forager because of a very peculiar mental tug or nag that food exerts on my brain.

Let me try to explain. Let's say that I am snacking on a piece of toast or a cracker with cheese, but then I get sidetracked by something—a telephone call perhaps, or a request from one of my daughters to help her find her book or her socks. I put the half-eaten morsel down on the table or a windowsill and seemingly forget about it. But sometime later, something inside my head won't let me rest until I've gone and retrieved that bit of food. I don't need to try to remember where I put it down. I usually walk straight to the spot, guided by this peculiar sensation of unfinished ingestive business.

I can't be the only one to experience this near-magnetic attraction toward half-eaten food. Everyone, I think, must be built this way. But none of the people to whom I tried describing this phenomenon seemed to recognize it. That is, until I spoke with Lewis Barker, a professor of psychology and neuroscience at Baylor University in Texas. I called Barker especially to ask him about this because I knew from his writings that he was interested in both food and memory and had, at one time, thought a lot about something he called the "Mom's apple pie is best" phenomenon. How, he wondered, had the taste, look, and texture of his mother's apple pie gotten into his memory in the first place, and how had it managed to persist there for decades? Barker, I know, would have had a great deal to talk about with Marcel Proust. And with Lin Yutang, the Chinese writer who said that patriotism is the memory of foods eaten in childhood.

"Sure, I've experienced that," Barker said during a telephone conversation in which I described my mental tug to him. Then we discussed how losing food could provoke this tug, but not losing our glasses or our keys. Those we had to find by a much more conscious reenactment of our movements—or if that failed, a thorough search of the house. What did Barker think was the reason for that tug, that mental honing in on lost food? He had always attributed it, he told me, to the fact that he was a tidy person and liked to finish what he started. But he also thought it might have to do with the fact that his parents grew up in the Depression and that when he was growing up, he was made to finish everything on his plate.

"But I'm a fairly messy person," I told him, "and I grew up with an indulgent mother."

He was quiet for a moment, and then he began to slowly muse. "Okay," he began again. "So there could be an entirely different slant to this. Like other animals, we humans have evolved in order to solve survival problems, and the most important of these problems is finding enough food to eat. Our brains have evolved to help us in this search. They are wired to sense food and to remember where to find food. And even though we've spent the last one thousand or five thousand years in a city, our brains haven't changed at all. They are still wired in the same way. We are still foraging animals."

You can really see this in children, Barker went on to say, because of their increased nutritional and calorie needs. From the time his children were very little, they had always known every-thing there was to know about the food in the kitchen and where it was stashed: the chocolates on the top shelf, the cookies in the cupboard. "A teenager doesn't know where the dishwasher is, but she knows exactly what's in the refrigerator," Barker observed. "She may not know where to find an auto parts store, but she knows every place to eat in town."

That night at dinner, I repeated parts of the conversation I had had with Barker, and though my own daughter resented the comment about the auto repair store, she agreed to a very informal test of the idea that children have an increased consciousness about food. I asked both her and my husband to write down on a piece of paper the foods that were in our refrigerator. Not the foods that were usually in our refrigerator, but the foods that they knew to be in there that very night. I thought my husband would have a fighting chance in this little test since he had put the gro-ceries away just the day before. But still, he listed only items that were routinely in the refrigerator—cheese, fruit, milk. My daughter listed the foods that were actually there—Brie cheese, blueberries, two percent milk, raspberry and cherry yogurt, packaged and bot-tled yeast, et cetera, et cetera.

The next time I talked to my older daughter, a teenager with a remarkable memory for the events of her early childhood, I also asked her what she remembered about the kitchen and its con-tents. She was puzzled because very little came to mind. Then, a day or two later, she called me from school with a rush of food memories. But they were outside memories, foraging memories,

memories of her and her sister collecting honeysuckle and clover blossoms, acorns, and hickory nuts and stashing them in their tree-house or in empty flowerpots. I could remember those stashes too, remembered cleaning out those pots in order to fill them in the summer and stopping to wonder at the strange assemblages inside. My daughter had also asked her friends at school about their early memories regarding food. "That's the game that we all played," she told me, "finding food."

It is more than a game for children in other parts of the world. Among the Alor of Indonesia, children are not given any food between their morning and evening meals, and they soon learn to forage for themselves by scraping food from cooking pots, raiding the fields for vegetables, and collecting insects that would be spurned by adults. Young Tallensi children in Ghana satisfy their hungry stomachs by eating toads and snakes, animals that are disgusting to older children and adults and that they too will spurn in time. Young Hadza in Tanzania, hunters and gatherers who live the way all humans lived before the advent of agriculture, are active foragers, collecting berries, the fruit of the baobab tree, and digging for tubers. These children make significant contributions to their families' resources.

It used to be thought that children only foraged for themselves in agricultural societies and that, in societies in which people lived by hunting and gathering wild resources, children relied on their parents for food until they were adolescents. But this was a mistaken view that arose among anthropologists when too much emphasis was placed on the studies of just one group of contemporary hunter-gatherers, the !Kung San of Africa's Kalahari Desert. Among the !Kung, children do not forage because of the nature of the !Kung resources and the long distances that !Kung women must travel to collect food. It is more efficient for them to stay in camp after they are weaned and crack mongondo nuts, the staple food of the !Kung. Studies of other hunter-gatherers, though, including the Hadza, have found that even very young children are active foragers. My daughters would have been happy growing up with the Hadza, though they might have objected to the fact that only boys are allowed to hunt with bows and arrows.

So this thing that Barker and I share, this enhanced food aware-ness on the part of children, is part of our mental makeup as humans, part of our survival package as we foraged for food. And it seems only natural that we would be made thus, with built-in tendencies to search for food, and only natural that traits that help in this search— a memory for food, for instance, where one is likely to find it, where one might have stored or left it—would be strongly selected for by natural selection. But only in the last twenty years or so have biolo-gists begun looking at memory, memory in any animal, in this way.

Biologists and laypeople alike had always wondered at the food-storing behavior of animals like the gray squirrel, which buries hundreds of acorns every fall, or the coal tit of Europe's coniferous forests, a small chickadeelike bird that stashes seeds in some eighty thousand different locations. But until fairly recently, observers had assumed that these animals used a sense of smell to find their hidden caches. For who could even imagine the kind of memory that would allow animals to remember so many different hiding spots? But most birds at least do not have a well-developed sense of smell. Plus people couldn't help noticing that food-storing birds and mammals don't sniff around for their nuts and seeds as one would expect if they were locating their caches on the basis of smell. They head straight for them, as one biologist notes, "as if they know exactly where they are."

Then in 1986 two biologists offered the first direct proof that some animals do locate their caches by memory, not smell. Marsh tits are small birds, also relatives of the American chickadee, that are common in the deciduous woods of England. There, they compulsively hoard seeds, nuts, and dead insects underneath bark, moss, and lichens. Ornithologist John Krebs and a colleague at the Edward Grey Institute of Field Ornithology in Oxford had the idea of labeling and keeping track of peanuts that they gave wild marsh tits to hide and of also labeling and keeping track of peanuts that they themselves hid in sites close to each of the birds' caches. What they found later is that the birds searched many of their own caches for food but none of the investigators', a result best explained by the birds' use of their memory, not their sense of smell, to find the peanuts. In the same year, investigators at Wellesley College found that gray squirrels rely on visual and spatial cues to locate the hundreds of acorns and nuts that they bury and retrieve each winter, but they use olfactory clues—smell—very little.

And if you think about it, tits and gray squirrels must be made this way. If these animals didn't use their personal memory of where they had cached their food, what would be the advantage to caching food? Any animal that could smell at least as well would have the same chance of finding the food. Storing food makes sense only if the hoarder has some advantage over all its competitors and is more likely than they are to retrieve the stored goods. And one thing that would give them this advantage would be a precise memory of the location of a stash. More recent studies have found that this kind of memory exists in many but not all food-storing animals. Nocturnal animals, like some mice and rats—animals that don't have a very well-developed sense of sight—do rely more on smell and less on spatial memory to locate their stashes.

The part of the brain where the food memories of squirrels and tits are stored is the hippocampus, as investigators have also recently found. The hippocampus is anatomically different in birds and mammals, but it seems to perform many of the same functions related to long-term memory. Lesions in the hippocampus don't disrupt a food-storing animal's ability to store food, just its ability to retrieve its caches. All animals have a hippocampus, but the hippocampus is bigger in food-storing mammals and birds, and bigger still in animals that hide their food in a large number of small separate caches. Gray squirrels, which make hundreds of different stashes each fall, have a bigger hippocampus than red squirrels, which store their food in just a few large, well-stocked larders.

The hippocampus, like all regions of the brain, is an expensive piece of equipment to run, so nature doesn't dole it out lightly, and she doesn't expand on it unless necessary. Black-capped chickadees, which store thousands of seeds for the winter, don't start out with an enlarged hippocampus. Its volume increases only after the chicks leave the nest and actually begin to store food. There is also evidence that the hippocampus expands and contracts in some animals during the course of the year, shrinking in size during the summer, when food is abundant, and expanding during the food-storing season.

As far as I know, no one has studied the hippocampus of foraging animals that do not hoard food but that do return to the same grasslands or trees to feed each year—animals like the migrating wildebeest of Africa's Serengeti Plain or chimpanzees and monkeys, which somehow know to gather at trees as soon as they are in fruit or flower. A sixty-year-old elephant matriarch is an encyclopedia of food facts on which the rest of her clan depends to get through seasons, droughts, and other disturbances in the environment. What does her hippocampus look like, and how does it compare with that in a much younger elephant? And what about human hunter-gatherers? How does the hippocampus of a Hazda or a !Kung compare with those in us who buy our food at a grocery store? Perhaps, in humans, this part of the brain will shrink over time, though as

Dr. Barker points out, finding food will always be important. Here in the United States, at the end of the millennium, it's not as difficult as it once was, but that could change very quickly.

During wars, famines, or any time of shortage, people turn— return—to foraging to fill their bellies. Many countries in Africa still have what they call their "hunger season," a time of year when inhabitants run short of the foods that they have grown and stored and must find wild things to put in the pot. At the end of the American Civil War, Southerners were helped in their search for food by Frederick Aldolphus Porcher's *Resources of the Southern Fields and Forests*, a book written expressly and hurriedly for that purpose. During the Irish potato famine, country people survived on "hedge nutrition," a knowledge of what leaves and berries could be combed from hedges and other wild places.

Some Irish families are ashamed of what they had to do during the famine to keep hunger at bay, and Marie Smyth, an Irish professor born in the early 1950s in the north of Ireland, observes that "to this day, in parts of rural Ireland, a 'respectable' family would prefer pie made with apples that were bought in a shop. A 'respectable' family would not wish to acknowledge that they too knew about foraging in the hedges and ditches for wild foods to eat." But others think it important to pass on those tricks of survival. Of her own family, Smyth remembers, "As a child walking in the Irish countryside, my mother taught me that hawthorn berries and leaves are edible, but rowan is not. Blackberries, fitches (vetch), dandelion are also edible; soup made from nettles is full of iron, the fruit of the black thorn—the sloe—is edible but slow to ripen and bitter before it is ripe."

Like Marie Smyth's family and foragers around the world, I too prefer to keep my food-finding equipment in good operating condition. And I also enjoy being reminded, by that insistent little tug, that I am a forager in my heart and in my genes; that we all were foragers, hunters and gatherers, until a change in the human way of life allowed many of us to put foraging behind us, leaving food production to others while we moved on into new and different realms of thought and activity. But there's another reason why we shouldn't be too quick to leave our foraging pasts behind. A very good reason and the thing that impels this book. For if human nature was forged during our existence as hunters and gatherers, as we know it must have been, then understanding how humans and other animals forage for food may better our understanding of human nature and of the way that humans, men and women, interact with each other and the rest of the world. Understanding that "finding food" has shaped our human selves and turned us into the people who we are today may help us to unravel some of the many mysteries surrounding food and the consumption of food.

It was a surprising road that my husband's question set me on, and it sometimes took me as far away as the Arctic or tropical forests in Central America. Like any forager, I have gone where the picking is good and have tried to fill my mind, and the pages of this book, with whatever I found that was tasty, nourishing, and available. Just as one berry patch leads to another and then, perhaps, to a clump of ferns or mineral-packed lamb's-quarters, so the ideas in this book follow each other, sometimes beginning with animals and ending with humans, and sometimes beginning with food and ending with the origins of power and relationships between husbands and wives, or with conerns about human population growth. For these things, too, have their roots in food, and the stomach really does, as French biologist Jean Henri Fabré once observed, "sway the world."

Three

DISCRIMINATING DINERS

Unless you have scooped up a pail full of pond water at one time or another and placed a drop of it on a microscope slide, you may never even have seen the tiny creatures that inhabit every pond, every ditch into which water finds its way. They're called copepods, and they are crustaceans, miniature freshwater relatives of lobsters, shrimp, and crabs. To the naked eye, they are just tiny white specks jerking through the water. Under the microscope, though, they reveal themselves to be a flamboyance of tufted antennae and dangling egg sacs, with legs like pine boughs and tails like fern fronds, all in the length of a millimeter or two, shorter than the span of an eyelash. One of the most common copepods, cyclops, has a single median eye.

Small as these marvelous creatures are, they are an important link in aquatic food chains. They feed on microscopic algae and bacteria and are fed on, in turn, by macroscopic carnivores, small but visible fish.

Now consider this: for years, decades, centuries, ever since the early 1700s when Antonie van Leeuwenhoek first rendered cope-pods visible with his newly invented microscopes (thus disturbing everyone's ideas about what constitutes "pure" and "clean" drink-ing water for the rest of time), copepods have been thought of as filter feeders, passive consumers of whatever life dished out to them in the way of pond water. And who would ever have thought oth-erwise of a predator less than the size of a pen tip and of prey as small and numerous as algae and bacteria? Who would ever have imagined that copepods pick and choose what they put into their mouths, sorting through clouds of dead algae, dirt, and dead bac-teria for those rare choice tidbits of live algae and live bacteria?

But in the 1970s high-speed films of copepods changed per-ceptions of how these animals live and eat (at least among those people who think about copepods). These films (at two hundred to five hundred frames per second) show that "the term 'filter feeder' grossly underestimates the sensory and particle-handling capability and behavioral complexity of copepods," as one biologist observes. They show copepods orienting their bodies to a particu-lar particle of algae, then changing the feeding currents of their antennae to increase their chances of actually capturing it. They show copepods tasting particles in their mouths before either accepting or rejecting them. They show that copepods are much more likely to reject low-quality particles like dead algae, plastic pellets, fecal pellets, and toxic algae and to accept living, nutri-tious algae.

Evidently, even the smallest consumers among us exhibit dietary discrimination. Even the smallest prey—algae and bacteria—have predators that stalk them individually. Nor is the gastronomic behavior of copepods unchanging or set in stone. Their fussiness about the foods they put in their mouths is not hardwired in their tiny brains. It is, rather, hunger-dependent, just as we would expect the selectivity of larger animals, animals such as ourselves, to be. In water that contains a low concentration of healthy algae, copepods discriminate only weakly against dead algae. In water that contains a high concentration of nutritious food, poor-quality food was greatly discriminated against.

And why should this come as any surprise when all animals have been shaped by natural selection to survive in the environments in which they live? Aren't we guilty of a certain kind of discrimination when we deem the copepod too small to be choosy about what it puts in its mouth? *Magnocentrism*, this might be called, the idea that only with a certain size comes behavioral sophistication and dietary discretion.

The small fish that feed on copepods make their own set of choices about what to eat and when to eat it. Hungry fish prey selectively on a species of very large copepod from which they are able to obtain a quick boost of energy in a very short amount of time. But after extended feeding, not-so-hungry fish prey selectively on a smaller species of copepod. This smaller species is harder for the fish to catch but easier to digest, *and* it contains many more fats or lipids, essential nutrients for fish as well as humans.

And it isn't just the nutritional quality of food that affects how an animal forages. When most animals forage for food, they also expose themselves to predators. If they are not vigilant and quick, they could be extending an invitation to others to dine. Animals must assume the risks of foraging in order to survive, but it's not surprising that they should minimize those risks as best they can. Gray squirrels, for instance, have to keep an eye out for cats, dogs, and hawks while searching for food. The safest way for them to feed would be to carry all the food they find under the protective cover of the trees before eating. But the most efficient way for them to feed would be to eat what they find where they find it. They resolve this conflict by striking a compromise between foraging efficiency and foraging safety. They eat the small things they find, the things they can eat quickly, the things it wouldn't be worthwhile to transport. And they carry the larger items—a pine cone, for example—back to safety.

And again, even very small animals try to minimize the chance that, in the act of eating, they too will be eaten. One of the ways that birds detect the presence of caterpillars is from the tattered leaves that caterpillars leave behind, in their feeding wake, so to speak. But as ecologists Bernd Heinrich and Scott Collins have found, some leaf-eating caterpillars screen themselves from the attention of birds (the gustatory intention) by minimizing this leaf damage. These caterpillars chew their leaves cleanly and leave no leaf tatters. Oftentimes they consume the entire leaf. Not all caterpillars exhibit these fastidious habits, only those that are palatable to birds. Unpalatable caterpillars, caterpillars that are poisonous to birds, have no reason to join the clean plate club, and they eat only the choicest part of the leaves. They leave the tough portions, such as the midribs and the large veins, and they leave the leaves in conspicuous shreds.

Natural selection, it seems, has been just as thorough in shaping the foraging behaviors of animals as it has their physical traits, their body size, skin color, enzyme systems. This doesn't mean that caterpillars or gray squirrels are conscious of eating in such a manner as to avoid the attention of predators. Rather, only those edible caterpillars that happen to feed in this way have survived and passed on their habits to their young. Only those gray squirrels that have struck their foraging compromise have been well enough nourished and have lived long enough to have offspring. Consciousness is not necessary for these extraordinary behaviors to develop—just the ruthless process of natural selection. I liken it to the effect that my style of weeding has on my garden. Because I am nearsighted, I tend to miss weeds that have the same leaf color as the plants that I am cultivating. So after several years, it now looks as if bittersweet is hiding out in the forsythia and barberry in the roses. How terribly clever of these plants to grow where they are least likely to be found and uprooted, one might think. But the cleverness is all in the selection process, my rather hit-and-miss weeding technique, which selects *against* weeds that don't resemble my plants and *for* weeds that do.

An animal's ability to forage effectively for food is one of its most necessary behaviors, so it must also be one of the most fertile grounds for natural selection. If food is scarce, animals that manage to nourish themselves better than their neighbors will be able to produce more offspring and their foraging techniques will spread throughout the population. Even if food is abundant, efficient foraging techniques will spread because the well-nourished, the quickly nourished, the safely nourished will have more time and energy for all the other things in life besides eating: mating, parenting, defending one's territory, resting. Even a predator like a lion, which spends most of its time resting, must be under strong selection pressure to hunt efficiently because hunting competes for the same limited time that all those other activities require.

As John Krebs puts it, "As long as a predator could improve its survival or reproductive success by hunting more efficiently, natural selection will favor efficient predators." And being an efficient predator involves not only such things as efficient teeth (teeth that can dispatch a prey quickly), an efficient heart (a heart that pumps enough blood to allow it to run as fast as its prey), and an efficient memory, but efficient behaviors as well—behaviors that enable it to avoid predators and select the best prey.

Though most of us have never spent much time thinking about the foraging techniques of animals, much less filming or observing them, this much most of us would willingly accept. This much makes some intuitive sense. But just how efficient is efficient? For more than a century after Darwin published his theory of natural selection, biologists were content with the idea that feeding must be efficient, but they were content also to leave it at that. What else could be said? Natural selection is ruthless, so feeding must be efficient. How nice. Was there anything else to be learned? Most scientists would have doubted it.

Then in 1966 an entirely new way of approaching feeding and foraging, and this question of efficiency, was born with the publication of two purely theoretical papers in the same issue of *American Naturalist*, one by John Merritt Emlen and the other by Robert MacArthur and Eric Pianka. What these authors proposed was simple but revolutionary. What if, they asked, we make the assumption that animals feed not just efficiently (we know that) but optimally, as well as they possibly can, given their physiological capabilities and the environment in which they live? One of the ways that animals might feed optimally would be to take in as many calories as they possibly can during the time that they spend foraging. So what if we make the assumption that animals forage in ways that maximize their net rate of energy gain over time? This assumption leads to certain definite predictions, and these predictions can be tested in the field or in the lab.

For instance, before the publication of these seminal papers, any reasonable person might have thought that an animal out foraging in the wild would pursue any food that it happened upon, any food that it recognized as food: a coyote stalking the grasslands of the American West would pursue any mouse or rabbit; a bluebird in my backyard would delight in any butterfly or worm. But the assumption that animals forage optimally generated some startling counterintuitive predictions. If animals are trying to maximize the amount of energy that they take in over time, these authors made mathematically clear, they will not eat just any food that they happen upon. Instead, they will develop feeding preferences that maximize the net caloric intake per individual prey. And they will *always* prefer foods that provide them with that maximum caloric intake, unless those foods are scarce.

In other words, if an animal is trying to maximize its energy gain, whether it preys on another animal is determined not by the population density of the latter—the chance of encounter—as most of us might have expected, but rather by whether that animal potentially provides enough calories to make the pursuit, capture, and consumption of it worthwhile. If it does, it will become one of the predator's preferred foods. If it doesn't, it will be included in the predator's diet only when preferred foods are hard to find. If preferred foods become abundant, it will again be dropped.

Optimal foraging theory, or OFT as it came to be abbreviated, immediately made sense of the behavior of animals in the wild, for it had often been observed that animals as diverse as barn owls, lions, and starfish are much more selective about their diets when food is abundant than when it is scarce. Lions during a drought, for instance, will prey on porcupines and bat-eared foxes, small animals that never catch their attention when antelope and giraffes are plentiful.

And it also made sense of some puzzling aspects of human diets. Why, for example, insects don't constitute a bigger part of human diets when they contain all the amino acids that humans need and the taste of some has been favorably compared to lobsters. When their capture entails so little risk. The reason, OFT made clear, is that they give so little bang for their buck. Unless insects are available in large quantities, or unless they can be easily collected—conditions that prevailed in the American West, where Indians *did* gather and eat huge quantities of grasshoppers and locusts, and in the southeastern highlands of Australia, where the seasonal dependence of aboriginal hunter-gatherers on the fatty, nutritious bodies of the bogong moth earned them the name of "moth hunters"—or unless other, more profitable foods are scarce, gatherers should ignore this food source because they would do better by pursuing other foods.

Or why the number of foods that are considered acceptable to eat is so much larger in poor, densely populated countries like China than in less populated parts of the world. Taste has much less to do with this than the fact that diets expand out of necessity and shrink when preferred foods are plentiful. Given the strange and varied foods that people do eat—squirrel brains in rural Kentucky and bats in Samoa—most of us probably think that humans eat just about anything. But around the world, human populations generally consume quite a limited range of the edible substances available to them. And optimal foraging theory helps us to understand how they make those choices. It's not surprising, for instance, that nuts and honey, two of the most energy-rich foods, are universally prized by humans. Or that pigs—the animal that humans eat the most of, the first animal to be domesticated by humans—are the most efficient of all the domesticated animals in converting food energy into protein and fat. Pigs convert thirty-five percent of their food energy, compared with thirteen percent for sheep and a mere six and a half percent for cattle.

Optimal foraging theory may not be able to tell us everything we want to know about food and the quest for food, but for the first time, it gave scientists a way of exploring the mind of the forager. In the years since publication of the two papers in *American Naturalist*, it has also received considerable experimental support.

John Goss-Custard, an English ornithologist, was one of the first to experimentally test the predictions of OFT in a study of how a certain kind of sandpiper—redshanks—forages for worms on the mudflats of European shores. Redshanks, Goss-Custard found, prefer the largest, most energetically profitable worms and eat small worms only when large worms are scarce. A more precisely controlled study of caged tits (chickadees), presented with large and small pieces of mealworm on a moving belt, had much the same results. When both large and small pieces were present at a low density, the birds were unselective in their eating. But when the number of large pieces was increased to a point where the tits could do better by ignoring small pieces, they became highly selective. Even when the experimenters kept the density of large pieces constant while they increased the density of small pieces until they were twice as common as large pieces, the birds still ignored the small but abundant prey.

In its simplest form, optimal foraging theory envisages foragers as setting out across a familiar landscape and encountering a number of different potential sources of food, some of which will be judged worth stopping for and many of which will be ignored in the expectation that something better will come along. The simplest measure of the reward for stopping to exploit a resource is its yield in energy, though it was always clear to Emlen, MacArthur, and Pianka that animals have nutritional needs other than calories and concerns other than food. Remember the fish that consumes a smaller species of copepod in order to obtain essential lipids? And the gray squirrel that has to juggle energy intake and safety every day of its foraging life?

In real life, animals don't always pursue the largest prey because the largest prey is dangerous and can fight back. Indeed, it has been observed that predators are pragmatists in that they pursue the most vulnerable animals, the young and the weak, when given a choice. In real life, foods can also be preferred or ignored because of the time it takes to find and catch them. Other studies, including some on humans, the most complex consumers of them all, have examined the effect of search time on prey choice. In a pioneering study of Cree hunter-gatherers in the coniferous forests of northern Ontario, Bruce Winterhalder used both historical and verbal reports to show that Cree hunters changed their quarry as they went from using paddled canoes, to motorized canoes, to snowmobiles. When Cree hunters had to paddle canoes in their search for food, small species like beavers, geese, and hares made very acceptable prey. But as they increased the speed with which they were able to move about the landscape, they also increased their selectivity about the prey that they considered it worthwhile to pursue. With the use of snowmobiles, hunters began concentrating almost exclusively on moose and caribou.

Similarly, in the forests of eastern Paraguay, the Ache, hunter-gatherers who had their first peaceful contact with outsiders in the 1960s, used to hunt with bows and arrows. Then they were happy to bring home a capuchin monkey or a string of small birds at the end of the day. With the introduction of shotguns, however, Ache hunters began spending more time in pursuit of white-lipped peccaries, a wild pig that is the largest game animal in the forest. Monkeys, small birds, and any of the hundreds of other potentially edible resources in the Paraguay forest are now disdained because they would lower these hunters' overall foraging returns.

And in real life, animals have many nutritional needs, and sometimes those needs aren't met by the diet that is highest in energy. The same sandpiper that makes an optimal choice of different-sized worms also prefers a smaller and less energetically profitable nematode to all other worms because of some as-yet-undefined nutrient that is only available in this prey. Moose in the north woods get most of the energy they need from eating the leaves of deciduous plants, but to meet their sodium needs, they spend precious time eating the low-energy leaves of aquatic plants. Pregnant moose have higher sodium requirements than bull moose or females that are not pregnant, and they spend even more time browsing on those salty, low-energy leaves.

Ache hunters increased their selectivity when guns were introduced into their culture, but anthropologists studying the foraging habits of these people have concluded that if hunters really wanted to maximize their energy returns, they should abandon hunting altogether and concentrate instead on gathering palm fiber, a task performed largely by Ache women. "Ache men could obtain approximately 2,630 calories per hour if they did nothing but exploit palms all day," say anthropologists Kim Hill, Hillard Kaplan, Kristen Hawkes, and Magdalena Hurtado, who spent five years living with the Ache. "Instead they hunt seven hours per day with mean caloric returns of only 1,340 calories per hour."

Why do they hunt then? Why did they ever begin to hunt, and why do they persist in this seemingly unprofitable activity? The foraging strategy of Ache males makes sense only if meat and plant resources are not equivalent in value to the Ache. Because of the protein and fat content of meat, two essential nutrients that humans require in some quantity and cannot get, or get as easily, from vegetable sources, and/or because meat, which is shared widely outside the family, gives men social and reproductive benefits, 1,340 calories derived from meat may be worth more to Ache hunters than 2,630 calories from palm fiber.

This doesn't mean that carbohydrates will always be valued less by humans than proteins and fat. In the far north, where carbohydrates are in short supply, Inuit people living on the coast harvest seaweed to provide themselves with energy-rich carbohydrates, as well as essential vitamins and minerals. They harvest it all year round through holes in the ice, and inland groups travel long distances to make willing though seemingly disadvantageous trades of caribou for seaweed.

Nor does this mean that the Inuit and the Ache or moose chest high in a bog are not foraging optimally. When Emlen, MacArthur, and Pianka picked energy as the coinage of their new theory, they understood that they were making a great oversimplification. But they wanted to start somewhere, and energy seemed like a good place since all animals need energy in order to grow and reproduce. Other variables—salt and protein requirements, predation concerns, the effects of learning and experience, search time, special dietary needs associated with pregnancy—could be added to the foraging equations later, when and if they were necessary. The scientists in this field are sometimes criticized for adding variable after variable to their equations to make the behavior of an animal fit the observed data, but these critics fail to see the point. These equations aren't simple, because life isn't simple and animals have different nutritional needs and different concerns. All animals must make trade-offs of various kinds. Optimal foraging theory gives us a way of understanding those needs, concerns, and trade-offs. As one scientist puts it, "We'll never know how animals forage unless we try these experiments, however fraught with difficulties they are."

Optimal foraging theory doesn't propose that natural selection favors the best *conceivable* foraging strategy—whatever that
is. Natural selection, after all, can only work with the raw material
that it is given. It can only select the best strategy among existing
strategies, and strategies are always changing as circumstances and
environments change.

I think about optimal foraging theory all the time
now when I am out in the woods looking for food.
How many blackberries or thimble-shaped wine berries
do I need to see before I consider it worthwhile to stop and pick
them? Why is it so difficult to strip a berry bush completely?
How many berries are left on a bush when I feel that strong urge
to move on? Even though I don't depend on the foods that
I gather for my survival, I often feel as if my actions are being influenced by something that is trying to maximize my picking rates.
That bush is no longer worth your time and effort, it seems to
tell me.

I also notice the difference in the amount of time that I am
willing to spend searching for different foods. For the fragrant
chanterelle or the earthy morel, I have almost endless time and
patience (as long as I know they are in season), but I will gather
garlic mustard, for instance—one of the few wild, edible plants in
my area that deer have not eaten into oblivion—only when there
are no greens in my garden. And even then, only when I see a
place where these low-growing, cold-adapted plants, whose leaves
make a pungent addition to soup, are growing very thickly and I
can pick a great deal of them in a short amount of time. *All right*,
my foraging policeman seems to say, *it's winter; there's nothing else*

to gather, so go ahead and pick garlic mustard. But pick it here, where you can pick a lot. I'm not saying that I have no control over my foraging actions, for certainly I could force myself to pick garlic mustard in the summer or all day every day, year round. But these urgings interest me. I feel they are whispered reminders of the competitive circumstances under which all life evolved.

Four

THE SURROGATE FORAGER

Ellen Dierenfeld, head nutritionist at the Bronx Zoo and director of the Zoo Nutrition Center, a consulting service that provides zoos around the world with nutritional expertise, doesn't concern herself with theories of optimal foraging, but she probably knows more about the specific nutritional needs of more different kinds of animals than any other single person. And what she doesn't know, she will find out. If you have hedgehogs that are growing fat on the food you are feeding them, or if your gorillas have arteriosclerosis, or you need to concoct a diet for a rare species of monkey or rodent that feeds only at night, Dierenfeld is the person to call.

Dierenfeld is something of a rare breed herself, and when I spoke with her, there were just eleven wildlife nutritionists in all the zoos of North America. This baffles her, since zoos have become more and more concerned with conservation and since nutrition, she believes, is the foundation of conservation. "With-out good nutrition, you don't have good health," she told me, "and you don't have reproduction. You don't need genetics research because you don't have any animals or any genes."

Nutrition is the foundation of conservation is not just Dierenfeld's belief. It is her operating principle, her mantra, her first commandment, and she illustrates it over and over again with stories of well-intentioned zoo personnel feeding their animals with foods that jeopardize their health.

In many zoos, alfalfa is the standard feed for all kinds of browsing animals, like the rhinoceros, camel, deer, and giraffe. But alfalfa, Dierenfeld says, is too rich for most of these animals. It is much too high in protein and therefore puts a strain on the kid-neys and liver. Akopis, close relatives of giraffes found only in the Congo basin, have a natural diet that is almost identical to alfalfa in protein content, but for the rest of the browsers, Dierenfeld now recommends a mixture of alfalfa and low-protein, higher-fiber grasses. For camels, she has found, even this mixture is too rich. "What camels eat in their normal habitat is worse than crabgrass. For them, straw hay [hay that has so little nutritional value that it is usually used just for bedding] is adequate if they are not work-ing or reproducing."

A diet too high in protein was also part of the reason why the zoo's hummingbird population failed to breed in the fifteen years before Dierenfeld arrived at the zoo. When she first arrived, it was explained to her that most birds simply do not do well in cap-tivity; she shouldn't expect them to breed successfully. Then she read a study that showed that hummingbirds have very low pro-tein requirements. "Surprise, surprise," she says, knocking her head with her fist to emphasize the ridiculousness of the idea that birds that feed on nectar, a liquid composed predominantly of simple sugars, could possibly have high protein requirements. She and the bird curator convinced the head vet to begin feeding the birds a product that duplicates the ingredients of nectar, and in a few weeks the hummers were laying eggs. Shortly after that, the eggs hatched out.

"It's so much fun when biology works," Dierenfeld grins somewhat mischievously. She is young and blond with an open, infectious laugh and a tomboyish look about her. It doesn't surprise me at all that she grew up on an Iowa farm or that her father was a vet. "Birds are great," she says, "because if you get something right, you see the results right away. Reptiles do everything so much more slowly that a change in their diet takes a long time to show results."

Vitamin E was also a problem for many birds in the zoo, for without adequate amounts of vitamin E, the piping muscles in their chicks degenerated, and the chicks were unable to break out of their shells. Soon after her arrival at the zoo, Dierenfeld also changed the amount of vitamin E that was given to the Congo peafowl and fixed their hatching problems. When she found that peregrine falcons in captivity had vitamin E levels ten times lower than those in the wild, she experimented with the grain that zookeepers fed to the falcon's favorite food, live quail, by supplementing it with vitamin E. Soon the hatching problems of the falcons were fixed too.

"A few little successes like these," Dierenfeld says, "really made an impact." Suddenly she didn't need to work as hard to convince people that the diets of animals in the zoo should reflect the diets of those same animals in the wild.

Vitamin E also turned out to be a problem for much of the zoo's mammalian population, particularly the endangered black rhino. "We were dumping incredible amounts of vitamin E into their diet as a supplement for their dried grasses, and it still wasn't enough." Then, on a special trip to Africa to study the rhino's natural habitat and diet, Dierenfeld collected the plants that the rhino foraged on and analyzed them for vitamin E. "It turns out that these fresh green plants have almost ten times the vitamin E of dried forages," Dierenfeld observes. "The vitamin E that we feed our agricultural animals may be adequate for livestock that live a year and then are turned into hamburger, but it's not enough for animals that we want to live longer and actually reproduce."

Dierenfeld tries to duplicate, not the actual foods that an animal eats in the wild, but rather the nutrients. Duplicating foods would be impractical and prohibitively expensive for a zoo, but duplicating nutrients can sometimes be very simple. Figs, for example, are eagerly sought after by numerous animals living in tropical ecosystems, and Dierenfeld wondered what it was about figs that made them so popular. The obvious answer was, their sugar or their protein content. But when she analyzed figs from many different sources, she found that these two ingredients vary greatly from one batch of figs to the next. What didn't vary was the fig's calcium content. All figs are extremely high in calcium, a mineral that all animals need for healthy bones (and healthy eggshells if the animal is one that lays eggs). Interestingly, the fruits that zoos normally feed to their animals—apples and bananas—are very low in calcium. Figs are too expensive to use as zoo feed, but Dierenfeld could buy calcium supplement very cheaply and add it to the diet, thereby mimicking a diet rich in figs.

Dierenfeld has been thinking about the issues of feeding animals properly ever since she was a young girl. "As a kid," she remembers, "it made me sad when animals died and it could have been prevented. We had bunnies for pets, and we overfed them. Bunnies should be fed just two times a day, because that's all their moms do. They feed them in the morning, and then they walk away. They let them sit there and digest, digest, digest. But when we were kids, we stuffed them full of milk. Then, when they'd get diarrhea, we'd just give them more milk. It was absolutely the wrong thing to do. I knew that they shouldn't die when we were doing everything we could to help them. And it made me mad, too, because my dad was a vet and he should have known better. But," she adds, "most vets, like many medical doctors, get very little formal nutrition training."

So she decided, a long time ago, that she'd rather keep animals healthy than try to fix them up once they were ailing. And instead of becoming a vet like her brother and her father before him, she began studying to become a fisheries and wildlife biologist. When she realized that she would probably end up spending her life checking hunting licenses, she switched to animal science and nutrition and then set out to combine her two interests.

"It was a time when zoos were just beginning to think about issues of nutrition and feeding behavior," Dierenfeld explains. "Wildlife nutrition used to just be about combating agricultural pests. [Once you know what an animal eats, you know how to poison it, she says bluntly.] But Jane Goodall and a few other field biologists got people interested in the food of wild animals. They spent years observing one species of animal or another and came to realize just how much of any animal's day was spent in feeding and in activities related to feeding."

That work changed the focus of zoo nutrition programs. Zoos used to just put food in an animal's cage, but now they're much more aware of how hard an animal must work to obtain that same amount of food in the wild—and how different a wild diet can be. Take the hedgehog, for example, an animal that is very prone to obesity when kept in captivity. In the wild, hedgehogs eat their body weight's worth of food every night, four to six hundred grams, but they eat primarily insects, with just a smattering of small vertebrates, plants, and eggs. In captivity, they were often given four hundred grams of canned cat or dog food and just the occasional insects and plants. As Dierenfeld explains, "Insects have shells or exoskeletons, so the natural diet of hedgehogs is much more dilute, much less digestible than the food that we've been giving them. Plus, in the wild, hedgehogs wander for miles every night to find their body weight's worth of bugs."

Today, at the Bronx Zoo, hedgehogs are provided with exercise wheels and lots of greens, for Dierenfeld has found that cellulose has a structure that is almost identical to the chitin in insect shells. Carnivores are fed entire animals or large bones along with their meat so that they too get their "fiber"—fur, claws, teeth, and bones—as well as the minerals that those indigestible parts provide. And apes and monkeys are given lots of browse and a high-fiber biscuit that Dierenfeld developed for primates. Before she arrived at the zoo, meat and eggs had been a regular part of the gorillas' diet, and the gorillas were suffering from the same things that humans suffer from on such a diet: obesity, arteriosclerosis, and heart attacks. Once those items were removed from the gorillas' diet, their health improved immediately.

"A big part of our problem is that we tend to overfeed everything," Dierenfeld observed one day when we were sitting in her tidy office above the zoo hospital, munching on delicious slices of dried grapefruit and strawberries from a bag of Creature Crunch, another high-fiber zoo product that she was developing with a food manufacturer. "We do it to ourselves. To our animals at home. To our animals at the zoo. Visitors do it by throwing food at the exhibits. It must," she mused, "be a part of human nature."

Five

FORAGING AT THE ENDS OF THE EARTH

"Why have you come here?" asked Anthony Gaston, a research scientist with the Canadian Wildlife Service. We were sitting drinking tea and swatting mosquitoes at a rough plywood table in the cook tent of his camp on Coats Island in Hudson Bay, fifteen hundred miles north of my home in New York and two hundred miles south of the Arctic Circle.

It was a strange question, given what I had just done to get to Coats, an island uninhabited except for a herd of caribou, a scattering of Arctic foxes, an unknown number of polar bears, and sixty thousand thick-billed murres—the small black-and-white seabirds that are the subject of Gaston's research.

First, there were the flights from New York to Montreal and from Montreal to Iqaluit on the southern tip of Baffin Island, a town of four thousand that was formerly known as Frobisher Bay.

In Iqaluit, I met up with Christine Eberl, Gaston's camp manager, and Luc Pelletier, a new French-speaking graduate student, and flew with them by Twin Otter to Coats Island, stopping along the way at Coral Harbor, a small Inuit settlement, to pick up Josiah Nakulak, a local resident who would be assisting in a survey of the island. When Josiah was a child, he and his family, along with four other Inuit families, had camped on Coats for a year, hunting seal in the winter and snow geese and caribou during the short summer, living on murre and murre eggs when game was hard to find.

The Twin Otter landed on a short beach at Coats, and within seconds the plane was full of mosquitoes. Outside, they were so thick, you could literally see the path that a moving body cut through them. It took the pilots only a few minutes to unload all of our gear and our five hundred pounds of supplies, and then they were gone, crushing mosquitoes against the windshield as they turned the plane back toward Baffin Island. All that was left was to haul everything up to Gaston's camp, situated on a high, flat shelf at the top of the cliff that rose sharply from one end of the beach.

Gaston, a trim, bearded Englishman in a brimmed hat and green oilskin jacket, met us at the bottom and suggested that we bring the personal gear and perishables up, then go back for the rest of the baggage when it had cooled off and after a cup of tea. It was about seventy degrees out, unusually warm for a place where the mean July temperature is in the forties, but because of the mosquitoes, we had to stay completely covered up as we zigzagged up the cliff, laden with backpacks and boxes full of vegetables, butter, cheese, and frozen meat.

The cliff was steep and covered with small white flowers (which would later ripen into cloudberries or baked-apple berries, I would learn, a favorite fruit of the Inuit) and mosses so deep and spongy that walking on them was like walking on a pile of wet diapers. Unlike diapers, though, these sodden masses released piles of mosquitoes with each step. So this was what permafrost was like: rock-hard, frozen ground that gave the melting ice no place to go. Though it was wet and soggy on this cliff, the Arctic, I had to remind myself, got less precipitation than the Mojave Desert.

At the top, after I met the last member of Gaston's research group, graduate student Mark Hipfner, and as we were sipping tea out of stained plastic mugs in the orange light of the cook tent, Gaston asked why I was there. I had, of course, written him about the book I was working on, so now I just muttered something about wanting to observe how murres feed and raise their young under the difficult conditions the Arctic imposed. I hoped that this would buy me some time as I took in my surroundings, for I really did feel as if I were teetering at the end of the earth.

"Oh," he shot back, "if I thought you had an incredibly-harsh-environment angle, I would have told you not to come. There is nothing harsh about these cliffs. The birds are very fat. So fat that they're hanging around the cliffs instead of feeding."

That was the extent of our first conversation and the first time I experienced the prickly tongue of this very bright and very quick-witted scientist, a whiz kid at Oxford, I had been told, one of several Oxford graduates who had become prominent figures in Canadian ornithology. I would have liked to disappear, but that was not an option. The Twin Otter would not be back for another ten days, and in the meantime I was totally dependent on Gaston. Could he be having misgivings about inviting me to Coats? I thought of our correspondence over the previous months and of the cautionary note that was always at the end of his letters, a note that seemed to increase in intensity as the time of my visit approached.

". . . the Govt of Canada will not accept any responsibility, except what common humanity dictates."

"Plane crashes, bear encounters, and falling off cliffs are all possibilities and should be carefully considered. You may want to review your medical insurance."

Now that I was in this remote camp, whose only connection to the outside world was a shortwave radio, I could appreciate his concerns. Not all graduate students take to camp life, I later learned, and here I was a woman writer in her forties, a completely unknown quantity. Now that I was on the island, I was, in fact, amazed that he had even considered taking on an extra person under these conditions where sleeping quarters were so tight (the six of us were to sleep in bunks in an eight-by-sixteen-foot cabin), and where every ounce of food and equipment had to be hauled up that cliff.

I only hoped that my back and my bug spray would hold out so that I could be useful during my stay and Gaston would not regret his original generous impulse.

Gaston camped at the top of the cliff, although it meant lugging gear up 240 feet, to avoid encounters with polar bears, those fearless, unpredictable, top-of-the-food-chain mammals that are actually classified as *marine* mammals because they are equally at home in the sea and on land. They move easily from landmass to landmass by swimming or hitching a ride on a floating ice pan. Polar bears were also the reason why the various buildings—the cabin, the cook tent, and the supply tent—were separate and well spaced. In his first summers at Coats, Gaston had camped on the beach and had brought along a dog as an early warning device, but over time he found the dogs to be much more trouble than they were worth. One time he forgot dog food for that year's canine and wound up feeding him murres, his research subjects, instead. Another year's dog balked at being left alone when the researchers departed for the day and barked constantly, bear or no bear.

So Gaston gave up dogs for the protection of the cliffs (bears are not likely to climb them unless they have a very good reason) and for guns and a commercial Mace-like product called Bear Scare. One of the first things we did when we arrived at camp was to learn where the Bear Scare and firearms were kept and how to load and shoot the guns. Anytime we left camp, even to go to the toilet (the biffy, as it was known in camp, a plywood construction cleverly situated over a deep ravine), we were to take some form of protection with us. But I had my doubts as to how effective any of this would be. "One of the few generalizations I can make about polar bears is that they always seem to do what you don't expect them to do" was one of Gaston's least reassuring comments. Walking along the beach was an entirely new experience there in the Arctic, where ice pans dotted the water and where an ice pan could be a convenient raft for a polar bear who could come out charging. It was definitely not a time for absentminded daydreaming. Sharks with legs, I thought, and I thought also of how lucky most of us are not to live with the gruesome possibility of being eaten alive—a daily possibility for most of the world's creatures.

But that evening, when Gaston's assistant Christine took the new graduate student and me down into the murre colony for the first time, I began to see why researchers would put up with the bears, the cliffs, the close quarters, and the mosquitoes and return to this island year after year. From the plateau where the camp was located, the murres were invisible to the eye, the ear, and the nose, but the minute we dropped over the side of the plateau, all that changed.

Suddenly we were bombarded with murre smells and murre sounds, and with the sight of murres flying furiously through the air, murres dotting the water, and murres occupying every square inch of rock, from the high-water mark all the way to the top of the steep cliffs. They looked like penguins with their black-and-white evening-dress plumage, but penguins in miniature. Penguins only fifteen inches high, with sharp, daggerlike bills for spearing fish. Some sat alone on tiny protuberances of rock; some, in long rows along long narrow ledges, like necklaces of murres, their black backs to the sun to absorb its warming rays, their white bellies pressed up against the rock. Others clustered on the wider ledges, in batches or in rows.

Descending into the colony was like walking into an open-air arena filled with hysterical, raspy-throated loonies.

Ha . . . . !   Ha . . . . !
Ha . . . . !   Ha . . . . !
                Ha . . . . !
Ha . . . . !
                                    Ha . . . . !

they snorted and snickered from every direction, though it was impossible to tell where one of these raucous calls ended and the next began, for each seemed to trail seamlessly into the next. This undulating score of madness was broken only by another sound the birds make, the sound that has given them their onomatopoeic name, pronounced, with a low growl, to rhyme with *fur*. Ha . . . ! Ha . . . ! Ha . . . ! murre . . . murre . . . murre . . . Ha . . . ! Ha . . . ! Ha . . . !

The birds that were sitting on the ledges were incubating eggs, Christine explained, and those that stood directly behind them were their mates. Males or females, it was impossible to say since the murre sexes are identical in almost every way, and the

only way to tell them apart is to watch them when they're copulat-
ing, then note who's on top. There were also many birds circling
the colony and attempting to land, and these, Christine said, were
trying to stake a claim to a piece of rock, to a place where they
could lay and incubate their single blue-green egg.

With all the soft moss around, it's hard to believe that the
murres would opt to lay their eggs on the bare rocks of the pink
cliffs (pink because of all the bird droppings). But it is the rocks
that give the murres a measure of safety from the Arctic fox,
whose agile reach determines exactly where murres can roost and
live to tell the tale. The cliffs don't provide safety from all predators, though, just land predators. Living there among the murres
were a dozen or so pairs of gulls who treated the colony as a
twenty-four-hour convenience store. Every so often one of these
gulls would fly past the colony, its head screwed to the left as it
searched the cliff for an unattended egg or, better still, a newly
hatched murre to feed to its own chicks. Each gull chick requires
three murre eggs or three murre chicks a day, a little less than a
pound of food. "Murre, murre, murre," a group of birds murmured
after a gull, scanning the cliff for unattended eggs, passed them by.

Christine took Luc and me to all the different blinds that
night, the four-foot-by-four-foot plywood boxes that Gaston and
his colleagues had hung from the sides of the cliffs in order to be
able to observe the birds. In the endless light of the Arctic summer, we followed well-worn paths in the thick mosses until the
cliffs became too steep for moss or grass, then lowered ourselves
down by ropes fastened to the rocks. Gaston hadn't been exaggerating about the possibility of accidents, I realized. Last year,
Christine told us, a large boulder went through the roof of one of
the blinds just minutes after a researcher had left it.

E arlier that night at dinner, over a beef curry he had cooked, Gaston had brought us up to date on all that had happened since he and Mark, the graduate student, arrived on the island in early June. The weather had been extremely warm (global warming is never far from the minds of Arctic researchers), and the ice had gone out early that year. It left the Coats shoreline at the end of June rather than in the middle of July. This was good news for the birds—they would not have to fly as far to find open water, where they could dive for food—but bad news for the scientists. The snowbanks in which they stored their perishable foods usually lasted all summer. This summer, though, those snowbanks would probably melt by sometime in July. The camp had an adequate supply of canned goods, but water was going to be a problem. The runoff from those snow-banks provided the camp's washing and drinking water, and already it had slowed down noticeably. The nearest alternative source of clean water was almost half a mile away.

Gaston, called Tony by everyone in the camp, also brought the new arrivals up to date on the birds. He and Mark had done several twenty-four-hour watches from inside the blinds and had found that the birds copulate at night and that the male and the female incubation shifts are just about equal: twelve hours and three minutes for the females and eleven hours fifty-seven minutes for the males. Thick-billed murres, it would seem from this one piece of information, have a remarkably egalitarian child-rearing arrangement. But their life is more complicated than that. When the chick is about three weeks old, it leaves the cliff in the com-pany of just its father, swimming with him toward the murre wintering waters off the coast of Newfoundland. The female stays behind on the cliffs, then flies to Newfoundland directly.

So now you might think that murre fathers put much more time and energy into their young than murre mothers. But Tony and his colleagues have found indications that the female's part of life's bargain—egg-laying plus her half of incubation, brooding, and feeding—is in fact the more onerous. When the birds first arrive at Coats from Newfoundland in early June, when the cliffs are still covered with ice and snow and the Arctic fox still wears its white winter coat, investigators rarely see the female birds, for they are off at sea fattening themselves up on cod and capelin so that they will be able to produce their one-hundred-gram eggs. Females, they have also learned, are much more likely than males to skip breeding one year, another indication of their more difficult lives.

While I was in camp, the researchers' days were spent mostly sitting in the blinds counting the number of birds that were incubating eggs. The messy part of their work would come later in the summer, when they would hang over the sides of the slippery cliffs by ropes and band the legs of one thousand newly hatched chicks. It is these bands, which transform identical-looking murres into individuals with their own futures and personalities, that enable the researchers to ask, and answer, questions about murre ecology and behavior. *At what age do murres begin to breed? Why are some murres more successful than others at raising chicks? What percentage of murres return to the cliffs each year, and what percentage are lost in Newfoundland, where some one million birds are shot each winter for food?* (More when Canada first closed its 350-year-old cod fishery because of overfishing, but far less once the limits recommended by Gaston were put in place.) And, of course, *How do murres divide up the business of child-rearing?* No one looks forward to fishing around for chicks on the sloppy, smelly ledges, to the possibility that gulls will bombard them from above or that chicks will fall off the ledge, or that they will step on one of last year's unhatched, uneaten eggs. ("When that happens," Mark told me, "it's like an explosion of nerve gas.") But banding is necessary to obtain the kind of information that scientists will need to help protect these birds as conditions in the Arctic change.

Since I didn't have research responsibilities, I helped out when I could or sat in an unoccupied blind and watched the birds on the cliff and the beluga whales and the walruses in the cold, clear water below. That's where the murres would be, too, if not for the fact that they need a dry place to lay their eggs. Murres are built for diving, not for flying. Their short wings enable them to flap their way three hundred feet below the surface of the sea to search for cod and capelin. But aboveground these same wings make them look like small flying pillows. They fly awkwardly, using their webbed feet as rudders, and if they leave a ledge, it takes them several minutes to maneuver their way back. They also land awkwardly with their heads bowed, as if embarrassed to be seen doing something so inelegant. "Murre . . . murre . . . murre," a ledgeful of birds chant in unison when a visiting murre, a prospector in search of a place to lay an egg, has been harassed into leaving the ledge. The interior of their mouths is the color of squash flowers, and as I sat in my plywood box, I pictured the verdant vegetable garden that I had left far behind.

Some of the birds were so close that I could have reached out the window of the blind and touched them. To sit there among them was to be inundated with questions. To be able to bring these questions to Tony, one of the world's foremost seabird biolo-gists, for his very dry and very apt explanations was better than having a private bedroom or a hot shower at the end of the day. The birds did all look exactly the same, but they behaved so differ-ently. Some were very easily scared off their eggs, while others would stare down even the boldest gull. One bird was trying to incubate a broken shell, and another, three eggs.

"I know that bird," Tony said when I asked him about it later. "At least there was a bird in that spot incubating three eggs two years ago. What a kleptomaniac."

Another time, when I was peppering Tony with questions about why one of the gulls stood around for a very long time with a murre egg in its mouth and why incubating murres seemed so reluc-tant to get off their eggs when their mates returned for the changing of the guard, he pulled at his neat white beard and gave this long, musing answer. "Gulls are dilatory in everything they do. I think it's the nature of predators. The act of predation has such consequences for them. They could get injured in the act. They could kill some-thing they're not supposed to—their own chick, for example. Murres, on the other hand, have no problem distinguishing their egg or chick from their food because a chick doesn't look anything like a fish. Their problem is resolving things with their mate.

"'I want to get on the egg,'" Tony imagined that a typical murre domestic quarrel might begin. "'Well, I don't want to get off it.'" This is an argument, he added, that takes place much more frequently here at Coats Island, where the colony is small and the birds are well fed. At larger murre colonies in the Arctic—the one on nearby Digges Island, for example, where 200,000 pairs of birds compete for food—"when one mate comes in from the sea, it gives one squawk and the other is off."

After observing the colony at Coats for ten years, Tony also has a very good appreciation for the differences in the ledges where the birds lay their eggs, for the fine points of these pink cliffs. One ledge tilts ever so slightly, and eggs have a tendency to roll off of it. Another leaves too much of the incubating bird exposed, and marauding gulls are able to pull it from the cliff. A third has no clear place for the chick to jump to the sea when it leaves the cliff. A fourth is exposed to the long reach of the Arctic fox. One night Mark came in with the news that bird number 30 on the P ledge had lost its egg, and Tony seemed to know exactly which bird he was talking about. "Well, it was a bloody stupid place to lay an egg" was all he said.

When Christine, Luc, Josiah, and I arrived at Coats in mid-July, no chicks had hatched as yet, but chicks were on everyone's minds. They always are. Their jump off the cliffs with their fathers is the most dramatic event in the life cycle of this bird—indeed, perhaps of any bird. Murres, as I've said, are not capable fliers, even as adults, but chicks jump from these 240-foot cliffs before their flight feathers have even come in. They can't fly, so they leave the cliffs by a combination of parachuting and gliding. On the night they jump (and most chicks jump together, in order, it is thought, to dilute the risk of their being nabbed by gulls or peregrine falcons), chicks move out to the edge of their ledges and flap their wings hard to glide forward. They jump out as far as they can, then drift downward. Those that fail to clear all the ledges are rarely injured in their falls, but they may be attacked by murres or gulls nesting below. What's worse, they may become separated from their fathers, yet they must have the protection and the guidance of their fathers in order to survive those many weeks at sea. The fathers leave the ledges with their chicks, and they try to adjust the speed of their descent to match that of their offspring, tilting back and forth to slow themselves down. But if their chick makes a premature landing, they are completely incapable of coming to a quick stop.

A murre chick's entire future hangs in this one jump, and chicks stare down long and hard before taking the plunge. "The chicks are real gamblers," says Tony. "They put everything they have on this one roll of the dice." But here at Coats, the dice are loaded—and not in the chicks' favor. Because the cliffs at Coats are not nearly as vertical as the cliffs at other murre colonies in the Arctic (at Digges, for instance), the chicks cannot clear them nearly as easily as they do elsewhere. At Coats, twenty percent of the chicks will die in this, their one and only chance at growing up. The first time Christine saw the carnage, she wept.

No one knows very much about what chicks do after they take to the sea with their fathers, but banded chicks are not seen back on the cliff for two years. The birds are fully grown at that point, but they do not begin to breed for another two years. Most do not breed successfully until they are seven or eight, when they have learned all that it takes to be a successful parent on these unforgiving cliffs. How to manage an egg on a narrow ledge slippery with bird droppings, for example. How to get along with your neighbors and your mate. How to find and hold on to a suitable place to lay an egg. This is one of a murre's most basic skills and one that it takes years to perfect. The colony at Coats is full of home seekers or prospectors. They circle the cliff repeatedly, landing on one ledge after another, where they are usually driven off by the squawks and jabs of that ledge's present occupants. Staking out a spot involves not only the appropriate level of murre aggression (some murres are so aggressive that they throw their mates off the ledge) but, even more fundamentally, the ability to find food efficiently.

Sitting in the blind at Coats day after day, watching these sixty thousand birds squabble over the opportunity to reproduce, I began to appreciate, really for the first time, how good nourishment is the basis of an animal's reproductive success. Murres that can find food quickly and efficiently are murres that can spend less time in the sea diving for cod and more time on the cliffs prospecting for a ledge. Murres that can find food quickly and efficiently are murres that are fit and strong, murres whose advances are harder for other murres to repel. Once one of these murres has secured a ledge, it can spend more time protecting it from other prospectors. It can be more vigilant against predators like gulls and falcons, vigilance that in other animals has been shown to decrease with decreasing body weight. Now it seems strange to me that I didn't have this appreciation all along. But food in America is so abundant that good nourishment is taken for granted and fat is something to struggle against, not embrace. Here at Coats, though, thin murres don't have thin offspring that bump along in life finding just enough food to get by on. Life is not that kind. Except in the most unusual of circumstances (when all the murres in a colony are food stressed and thin), thin murres will not have any offspring at all; they will not even be able to claim a ledge on which to lay an egg.

And here at Coats, good nourishment is the basis not just for an individual murre's success but for the success of the colony as a whole. At Coats, where competition for food is much less than at the large murre colonies, heavier birds lay heavier eggs that hatch into heavier chicks that leave the cliffs as heavier fledglings (or jumplings, as they are also humorously called) and return to breed at an earlier age than do their slimmer compatriots at the large colonies. The chicks at Coats are twice the size of the chicks at Digges when they leave the cliffs, and it is the size of the chicks, Tony thinks, that enables the Coats colony to thrive despite the chicks' high losses on the night they jump to the sea.

When Tony asked me why I had come to Coats, I wasn't really sure. I wanted to travel to the Arctic, of course, and observe the murres. What I got out of Coats, though, was this simple realization, so obvious now that it hardly seems to need expression, or my own expression, since I am hardly the first to understand this important connection.

"The whole of nature," noted the twentieth-century English scholar William Ralph Inge, ". . . is a conjugation of the verb to eat."

"Every living organism is just an abstraction when viewed apart from its physical and biological environment, for life is a continuous exchange of matter and energy," as the biologist Claude-Marcel Hladik once put it.

"One cannot think well, love well, sleep well if one has not dined well," Virginia Woolf observes in her book advising women to have a room (and income) of their own.

"An empty stomach cannot reason" used to be a saying common in Italy, a country where hunger was so prevalent until recent decades that a popular longing was to die of overeating.

Certainly I had heard these sentiments before, but somehow they hadn't meant very much until those days spent watching the murres' constant jousting over ledge space. Somehow I had managed to convince myself that life is a little gentler. Somehow I had been fooled into thinking that food and the quest for food are less important than they are, that life is not so much a struggle for existence as a profusion of resources. But as the murres began to show me, it is food that makes the world go round.

Six

THE HUNGRIER SEX

We always heard that it was love, but food makes love possible, as anthropologist Colin Turnbull reveals in *The Mountain People*, his terrifying account of a Ugandan tribe that was prevented from hunting on their traditional lands and forced to forage and farm in nearby barren mountains instead. In less than three generations, starvation had turned the Ik into a heartless people whose only goal was individual survival and whose only value system was food, or the possession of food—*the individual possession of food*. A good man, among the Ik, was not a man who treated his wife and children well, a man who helped to build a new school, a man who was a wise leader, but a man with a full belly. Among the many horrors that Turnbull saw during the years he spent with this people, 1964 to 1967, were young children snatching food out of the mouths of old people and a mother who rejoiced when her infant was taken by a leopard. She was happy not only because she no longer had to feed the infant but because she now knew there was a leopard in the vicinity, a slow-moving leopard because of its full stomach.

"Where no food nor receptacle is, even the Furies abandon the places," observed the sixteenth-century writer John Florio. But the Ik had no place else to go, and they abandoned instead many of the things we take for granted about being human, *many of the things we never thought could be abandoned*: love, cooperation, joy, empathy, the sharing of food. By the time Turnbull, curator of African ethnology at the American Museum of Natural History in New York, lived with the Ik, they had only the ghostliest remains of family or social structure. They still lived in groups and they still acknowledged family relationships, but men and women foraged and hunted for themselves and shared very little with either each other or with their children. After being weaned at about the age of three, children were expected to fend for themselves. Before Turnbull published his study of the Ik, few would have thought it possible that humans could live without customs of cooperation and food sharing, and among those who were the most shocked by his observations were anthropologists who had lived with other African tribes and had seen how extensively food was distributed among any group.

No doubt many blamed the transformation of this people on some character flaw in the Ik themselves, but as a proverb from the Virgin Islands says, "Full belly can afford to tell empty stomach keep heart." During World War II, when England was at the height of its wartime food shortages, M.F.K. Fisher published *How to Cook a Wolf*, a book of recipes geared to bring solace in the absence of butter and eggs and with a credo that "one of the most dignified ways we are capable of, to assert and then reassert our dignity in the face of poverty and war's fears and pains, is to nourish ourselves with all possible skill, delicacy, and ever-increasing enjoyment." It was a brave, fortifying credo and probably did the English much good, but their deprivations lasted for years, not decades, and they understood that the deprivations were necessary in order to maintain their society. The Ik were long past the point where advice or recipes could have helped.

There is a lesson here that we would rather not learn. Family and social structure are not things that can be taken for granted. The way that animals, including humans, interact is greatly affected by environmental conditions, and chief among these is the food supply. In the Kalahari Desert, female lions usually live and hunt together in prides, but when drought makes food scarce and lions must prey on smaller and smaller animals, prides disband and lions live a solitary existence. When circumstances improve, they reconvene. Groups of human hunter-gatherers in the Kalahari, and in other resource-poor areas, go through similar fluctuations in group size. Bands of !Kung and Hadza disperse into foraging units no larger than a nuclear family, both seasonally and in times of stress.

Lack of food can cause groups to break up, but abundant food can turn solitary animals into partygoers. Grizzly bears are usually loners and extremely intolerant of other bears. But once a year at McNeil River in Alaska, when salmon are so thick in the water that they nearly spill out onto the banks, sixty bears may come together in one place to fish. Large male bears that would normally confront each other stand side by side enjoying the nutritional highlight, the Thanksgiving Day, of their year. Bears arrive thin and put on weight rapidly, consuming as many as fifteen fish a day.

And food can also cause striking changes in the interactions between parents and their young. Murres are usually devoted parents to their single chick, working around the clock to brood it and bring it food. But if a cold spring has kept the cliffs locked in ice and the murres must fly unusually long distances to open water where they can dive for fish, they may abandon their eggs before they hatch and try again the next year. And what is true of murres has often been true of humans—hard as this is for us to accept. Records from the town of Limoges in France during the late eigh-teenth century reveal a close correlation between the number of children abandoned in any one year and the price of rye, the area's most important grain. In the event of a hunter's death, Inuit cus-tom used to require that his widow would kill any of her children under the age of three. We tend to think of behaviors like child care as being automatic and instinctive, that a murre with an egg, a parent with a child, would do its best to brood that egg and raise that child no matter what, but as one psychologist observes, "The enormous body of data that has appeared in the last several decades makes it clear that virtually no aspect of adult behavior is a fixed characteristic of a given species. They all vary as a complex function of key environmental conditions of which feeding ecology is one of the most important parts."

This recognition, that changes in food supply can change the sociability of animals and the way that animals care for their young, is really only a small part of a much more basic recognition: the importance of food in shap-ing every aspect of an animal's life. Scarcity of food can alter an animal's behavior, but food is also key to how that animal behaves under what we might call "normal" circumstances (though circum-stances, of course, are always changing). How an animal makes a living, as one biologist puts it, in large part determines everything else about that animal. Our lives are a reflection of the foods that we eat and how they are distributed in time and space. We truly are what we eat, as scientists have been finding out over and over again and in many different ways.

Some of these conclusions are not very new. "For it is difference in feeding habits that make some animals live in herds and others scattered about," Aristotle observed in the *Politics*, written some 2,300 years ago. Species of animals that are very picky about what they eat—animals like the klipspringer of Africa, which feeds on the fresh growth of small shrubs—must search large distances for their preferred foods and generally live alone or in small groups. Less selective foragers—the wildebeest, zebra, and topi, which dine on the abundant grasslands of Africa's Serengeti Plain—live in large herds.

And some of the conclusions are very familiar. It is well known, for example, that litter size in many animals changes with food supply and that populations of herbivores, which feed on the earth's abundant plant matter, tend to be much larger than populations of carnivores. Similarly, herbivores that eat leaves are more abundant than herbivores that eat flowers and/or fruit because leaves, the nonreproductive parts of plants, are much more abundant than the reproductive parts.

Some of them are also very intuitive, such as the relationship between the size of an animal's territory and the density of its prey. In the forests and brushfields of the central United States, where cottontail rabbits and muskrats are common, a bobcat may require only one or two square miles of territory to provide it with all the food that it needs. Farther north, where rabbits are scarce and bobcats feed instead on voles (of which they need to catch about twenty-four a day), one bobcat may require about ten square miles to meet its needs. In the mountains of Idaho, where life is even harder, it may require thirty.

The size of an animal's territory, as an optimal foraging theorist would reason, should be large enough to allow it to maximize its energy intake, but not so large that it would waste energy defending unnecessary turf. Three biologists found an ingenuous way to test this idea in a study of the very territorial rufous hummingbird, a small bird, but one known to drive butterflies, other hummers, and even bees from the area it controls. The biologists covered half of the flowers in a hummingbird's territory with cloth, then watched and monitored as the resident hummer increased its territory to compensate for the covered flowers. Territory size is tied to the availability of nectar, the biologists concluded, and hummingbirds can in some accurate way assess the amount of nectar they control and have available to them.

So territories or home ranges expand and contract with drought and whatever else affects prey densities. When biologists Mark and Delia Owens first went to the Kalahari Desert to study the behavior of the lions living in this inhospitable part of Africa, they had been told that during the dry season, when prey is diffi-cult to find, Kalahari lions migrate to different locations. What they discovered instead, as they tagged numerous lions, then tracked them in their truck and airplane, was that the lions just kept expanding and expanding their territories to meet their dietary needs. (And prides disband, as I've already pointed out, and lions include smaller and smaller animals in their diet.)

Far less familiar or intuitive, though, are the relationships between food and mating behavior and the different relationships that males and females have with food. For our understanding of these relationships is as new as Aristotle's observation is old. Only in the last twenty-five years or so have biologists realized that food might affect the behavior of males and females very differently. Only in the last twenty-five years have biologists even begun to think in terms of sex-related differences in diet and of how differ-ently males and females adapt to the same environment.

To understand why males and females would have different relationships to food and why biologists in the field must now frame their research in terms of these differences, one must first understand what it means to be a female or a male and the far-reaching consequences of what seems at first to be a very small difference in physiology. To be a female means to produce a small number of large, energy-rich eggs or gametes, sex cells that can nourish a developing embryo. To be a male means to produce a much larger number of smaller sperm or gametes, sex cells with no nutritive value. In all animal and plant species, the male produces sex cells that are much tinier and much more mobile and numerous than those of the female, a fact of life that is probably no accident. "An undifferentiated system of sex cells seems highly unstable," says the evolutionary biologist Robert Trivers. So as soon as selection favored those that invested their sex cells with nutritious substances, it also favored those that cheated the system and became adept at numbers and mobility instead. As soon as selection favored eggs, it also favored sperm. And there you have it: the origin of the sexes.

Females have drawn the short straw in this business of reproduction, and the consequences are endless. A male and a female mate; fertilization takes place. But because of the larger size of her sex cell, the female already has more invested in that fertilized egg than her partner. And because she invests more initially, she is the one more likely to continue to invest. She is the one more likely to incubate and protect her egg and to mind the offspring once they are hatched or born. She is the one more likely to evolve special equipment for the job of parenting, a uterus, egg sac, and/or mammary glands that secrete milk. You might think that, because it takes a male almost as much energy to produce all the many sperm cells in one ejaculation as it does a female to produce her one big nutritious egg, the two sexes are even on this score. But the difference is—and this is the difference from which all other differences between the sexes flow—that the female puts more energy into the one egg that gets fertilized than the male puts into the one sperm that does the fertilizing.

According to Darwin's theory of natural selection—the idea that life on earth is shaped, in effect, by the forces of competitive reproduction—all animals have been selected to have as many off-spring as they can successfully bear and raise.* This applies to both males and females, but because females produce their larger eggs more slowly than males produce their smaller sperm, males are capable of fathering more progeny than females can mother. A male's reproductive success, therefore, often depends on and is limited by the number of breeding females to which he can gain access. A female's reproductive success, on the other hand, rarely depends on the number of males she can mate with because only a very few matings will provide her with all the fertilized eggs she can possibly raise. Instead, a female's reproductive success is often limited by her ability to find enough food to nourish both her offspring and herself.

---

*It isn't that animals consciously carry out this Darwinian goal. But because the genes of animals that raise just a few offspring, when circumstances would permit more, are quickly swamped by the genes of neighbors that raise as many offspring as they can, competitive reproduction has become the way of all living things. Caring for offspring after they are born, of course, limits the number of offspring a parent can have, but it greatly increases the chances of their survival. So care, despite its disadvantages, has evolved in many different species. It is most likely to evolve in more predictable environments where the care that a parent gives its offspring has a better chance of paying off—a coral reef, for example, rather than the open ocean. In more variable environments, animals play by a different strategy. They produce thousands of eggs, with the expectation that a few will have the luck and genetic right stuff to survive.

It is not that males don't need food to carry out their repro-ductive strategies. On the contrary, males, being oftentimes larger than females, sometimes require an absolute greater amount of food. Nor is male fertility unaffected by food supplies, for studies in many different animals have shown that sperm count and motil-ity are reduced by inadequate caloric intake. Simply, as life evolved, the efficient collection and use of food, while obviously important to males, is not quite as important as it is to females. All other things being equal, as one biologist puts it, females spend most of their time looking for food, and males spend most of their time looking for females. Think of it this way: a male has to be strong enough to compete with other males for mates, but the male that spends too much of his time foraging and not enough competing may miss out on mating altogether. A female that spends more time than necessary foraging will rarely miss out on mating and may in fact do her offspring good.

This startling yet commonsensical idea is supported by reports on animals in the wild. In a number of primate and grazing animals, adult females have been observed to spend much more time feeding and searching for food than adult males and to show much faster rates of tooth wear. It even seems to apply to such heroes of feminist thought as the sea horse, a type of fish in which males incubate their mates' eggs in special brood pouches on their abdomens. Sea horse fathers were long thought to do most of the work of raising their offspring, to be "submarine saints," as Natalie Angier writes in the New York Times. That is, until recently, when a researcher at Amherst College found that, although the male does carry the developing brood, he does not nourish it. All the nourishment that sea horse embryos need to grow and develop is in their egg, nourishment that they receive from their mothers. In fact, by incubating these eggs, a sea horse father is probably less concerned with their development (or with being a help to his mate) than with freeing his mate up so that she can spend the necessary time feeding in order to produce more eggs for him to fertilize.

Like the authors of such popular books as *Men are from Mars, Women are from Venus*, the Swedish playwright August Strindberg believed that men and women are so different as to be of "two different species." Many scientists now are of much the same opinion. "One can, in effect, treat the sexes as if they were different species," observes Trivers, one of the first scientists to understand the profound effects on behavior that stem from this difference in the sizes of male and female sex cells. "Female 'species,'" he says, "usually differ from male 'species' in that females compete among themselves for such resources as food but not for members of the opposite sex, whereas males ultimately compete only for members of the opposite sex, all other forms of competition being important only in so far as they affect this ultimate competition."

Not long after Trivers published these very new ideas in 1972, in a paper in *Sexual Selection and the Descent of Man*, other biologists realized that these differences between males and females must have important social consequences. Before Trivers's influential paper, most people had been inclined to think that animals live in groups for protection against predators or because groups make it easier to find food. But Trivers opened the way to a very different understanding of social life.

Since females are the limiting resource of males and compete with other females to maximize their food intake, while males compete with males to maximize their access to females, it should be possible, Harvard biologist Richard Wrangham pointed out in 1979, to explain the distribution of females in the environment—

and ultimately of males—using the theory of optimal foraging strategies.

If, for example, a female's best strategy for finding enough food to eat is to maintain a territory from which she excludes all other females, but a male is unable to control more than one of those territories against other males, then the male's best chance at reproductive success, his only chance at reproductive success, may be to settle down with that one female and raise a family. The female might like to exclude males as well as females from her territory (except for a brief mating period) since males also compete with her for food. But males cannot be avoided entirely because they are often larger than females and because they are unencumbered by young. They are stronger, in other words, and can spend more of their time and energy trespassing into a female's space. In these circumstances, the female's best shot at reproductive success is to accept the presence of a single competitor who, if she chooses well, can exclude all other males and who, perhaps, can help her out with child care. This, Wrangham points out, is the strategy of gibbons and siamangs, apes that live in the forests of India and Indonesia in solitary monogamous pairs. Male gibbons and siamangs would be perfectly willing to accept more than one female into their territories, a willingness that biologists in the field have observed, but females won't let this happen.

That is one type of social and mating system. But what if food resources are so spread out that a female needs more space than she can effectively defend against other females? This leads to a very different kind of setup, Wrangham explains, one in which a female's best strategy is to eat as quickly as possible and to tolerate the presence of other females with whom she can cooperate in establishing a much larger territory. In this situation, one male wouldn't be able to exclude all other males from the territory, so males also have to tolerate the presence of other males. But they would, of course, continue to compete with those other males to gain priority access to fertile females.

This second scenario pretty well represents the social system of chimpanzees, species name *Pan troglodytes*, and it comes with many interesting twists and turns, including much female promiscuity. The territory required by chimpanzees is so large that the success of a female (her ability to successfully raise her offspring) depends on the males in her group cooperating to defend the territory against other troops of chimpanzees. Anything that a female can do to enhance that cooperation is to her advantage. Rather than carefully selecting one male to mate with, as gibbon and siamang females do, chimpanzee females mate with most of the males in the group, thereby increasing their common interest. And rather than quietly advertising their fertility to one male, as monogamous females do, chimpanzee females use their prominent sexual swellings to broadcast it to the entire troop.

These are just a few examples, but Wrangham was also able to make sense of the mating habits and social systems of gorillas and orangutans, and biologist P. J. Jarman applied these same principles and ideas to different species of African antelopes. Social systems and mating systems stem from fundamental differences between males and females, these investigators made clear. Females that live in groups have good reason for their behavior. Females that are promiscuous have good reason for theirs. Males that are monogamous are monogamous not by choice but because they have few additional mating opportunities—and/or their attentions are necessary to the survival of their offspring. For even where the distribution of females or resources provides the potential for harems or polygamy, males cannot always take advantage of it because raising an offspring may require both paternal and maternal care. On the cliffs of Coats, for instance, it would do a male murre no good to monopolize a ledge and acquire more than one mate, since it takes two parents working around the clock to raise just one chick.

No one knows what shape the foraging and mating strategies of the first humans—males and females—took or what the first human societies were like. Modern men have been called ecologically monogamous because they tend toward monogamy in some parts of the world, but take more than one wife in places where resources are abundant and where men, or some men at least, can accumulate or defend enough resources to support more than one family. But all males that are monogamous are ecologically monogamous. All are looking to maximize their reproductive fitness and will change their mating systems if circumstances allow it. An important difference between humans and other animals is that most animals are adapted to live in one kind of environment only: one type of habitat, where only one type of mating behavior is usually expressed. Humans live all over the world and in all different kinds of environments.

So muse on this the next time you find yourself thinking that human culture and human intelligence far remove us from the behavior and mating constraints of other animals. Or that women are congenitally monogamous and males incurably polygamous. In Tibet, on the southern arc of the Tibetan plateau, a place where all the arable land has long since passed into family holdings and where many of these holdings are so small that they barely suffice to support a family, one woman will sometimes marry two brothers, and together the three of them will work to raise their children. Polyandry—one female paired to more than one male—is extremely rare in mammals since females, in general, have fewer advantages in having more than one mate. But here it is, in our own very adaptable species. Polyandry is not the only form of marriage in Tibet. Nor should we expect that it would be. It is common only among the poorest people, those with the fewest resources or the smallest holdings. Most marriages in the area are monogamous, but rich landowners frequently take more than one wife.

Social and mating systems are some of the fallout from this essential difference between males and females, the fact that sperm are smaller and less costly than eggs. Males and females can also differ in size, shape, life span, growth and metabolic rates, and many other aspects of physiology, biochemistry, and behavior, including, as already mentioned, their eating behavior.

Adult females in a number of species have been observed to spend much more time feeding than adult males, and they have also been observed to eat different things. At the Gombe National Park in Tanzania, where Jane Goodall and her colleagues have been observing chimpanzees since the early 1960s, females exploit termites and ants twice as heavily as males, but males hunt monkeys and bush pigs much more than females. In another population of chimpanzees in the Ivory Coast, males and females differ in the amount of nuts they eat and in their use of stone hammers to crack those nuts open. Females crack many more nuts than males, and they crack them whenever and wherever nuts are available— on the ground or high up in the branches of a tree. Males, on the other hand, are much less proficient at the difficult task of opening nuts aloft, and they prefer to crack nuts at the end of the season, when they are dry and brittle and easy to open.

In the past, these feeding differences would have been chalked up to a kind of cooperation between the sexes, a way that males and females could divide up the resources of their territories. But that kind of cooperation makes biological sense (that is, it is only expected to evolve) only in a monogamous species, or a species in which a male maintains exclusive access to a group of females—in a situation, in other words, in which the resources that a male passes up would benefit his offspring or kin. The now extinct huia bird of New Zealand, a bird with a monogamous breeding system, had just such a cooperative feeding system, leading to great specialization in the huia bill structure. Male huias had short strong beaks for breaking wood in search of insects, and females had curved slender beaks for exploring crevices.

Species where males don't maintain exclusive rights to females can also show sex-linked differences in feeding behavior, but they must have other reasons. In these species, feeding differences may be techniques that allow males to get close to females without driving the females away. If males are not competing for the same resources as females, if they take on larger and more dangerous prey, or less accessible leaves and fruit, females may be more tolerant of their presence. On the Serengeti Plain, where giraffes live in fluid, open herds, male and female giraffes are distinguishable because of their different grazing habits. Males tend to stretch their necks to feed on the tops of the taller trees while females concentrate on the smaller, more reachable trees.

So some differences in eating habits are a form of cooperation between the sexes; some give males greater access to females; but others reflect the different dietary needs of males and females, the fact that successful egg-laying or pregnancy and lactation require an enhanced intake of calories and nutrients. Chimpanzee females, for instance, are probably better at cracking nuts in different situations and at all times of the year, because pregnant females (and their dependent offspring) need higher levels of protein and fat in their diet than fully grown males. For the same reason, females of many different species have been observed to feed longer than males—and in those species of animals that store or hoard food, females hoard more food.

Food hoarding in many species is predominantly a female business, something that I was reminded of the other day by a kindergarten teacher who was concerned that the girls in her class, but none of the boys, were hoarding the snacks that they were given in their cubbies. The teacher feared that this behavior might be an early sign of a troubled relationship with food. But would it change her perception to know that in bees, wasps, and ants, only females provision nests for their young and only females provision their own nests or hives for the winter? Or that female hamsters store twice as much food as males and biologists have found that permitting captive hamsters to accumulate food in their nests significantly increases the likelihood that they will show maternal behavior to a group of foster pups put in their cage? In hamsters, it seems, a well-stocked larder is a prerequisite for mothering instincts to emerge. What does it mean when a young girl hoards food?

Some animals store food, not in larders, but as fat on their bodies, and a few animals—humans included—do both. Large amounts of fat accumulation isn't an option for most animals because the dangers of being heavy and slow on one's feet outweigh the advantages of an energy cushion. So in most animals, the primary function of fat deposits is insulation from cold. But those animals that do have the ability to store energy in the form of fat also show differences related to sex. Both male and female bears put on fat during the summer to carry them over the winter, but female bears put on more than males. Both men and women have stores of adipose tissue (humans, by the way, are among the fattest of mammals), but on average, for young adults in an affluent society, fat stores account for twenty-seven percent of the body weight of females, but only fifteen percent of that of males.

"Women have brains, hearts, guts, skin, bones, and even saliva with physiological characteristics that differ from those in men," notes a Columbia University researcher. But one of the ways that human males and females differ the most is in the amount of fat tissue they carry. As anthropologists Peter Brown and Melvin Konner point out in a paper on obesity in the *Annals of the New York Academy of Sciences*, adult men are, on average, eight percent larger in stature than women and twenty percent larger in total body mass, but they have almost fifty percent less subcutaneous body fat as measured by skinfold thickness. And while there is some variation in fat distribution in different populations, Brown and Konner find that overall this dimorphism in fat tissue appears to be universal. Though the !Kung are very small in stature and extremely lean by worldwide standards, and the men are taller and heavier than the women, !Kung women have eighty percent more peripheral fat on their bodies than men. (Even this additional fat, though, does not keep !Kung women from worrying about food. Anthropologists living among the !Kung, a people that used to be stressed for food for part of every year, have noticed that women are much more likely to talk about food anxieties than men, about not having enough food to eat and about who did or did not give them food. Lorna Marshall, who lived among the !Kung in the Nyae Nyae region of the Kalahari during several periods in the 1950s, described an instance where she saw a woman go into a kind of semitrance and was heard repeating, for perhaps half an hour, that a certain hunter had not given her as much meat as was her due. "It was not said like an accusation," noted Marshall. "It was said as though he were not there. I had the eerie feeling that I was present in someone else's dream. [The hunter] did not argue or oppose her. He continued doing what he was doing and let her go on.")

It is clear, from the course of their development, that these differences in body fat have to do with reproduction and women's role in gestating and nursing their children. A difference in soft tissues is present in childhood (even little girls have more fat on their bodies than little boys), but it increases markedly during adolescence due to a great divergence in the rate of fat accumulation in adolescent boys and girls. It increases, in other words, at the time of reproductive maturation. The body fat of an average woman—thirty-four pounds on a woman weighing one hundred and thirty pounds—would go a long way toward sustaining a pregnancy or maintaining lactation in the event of a food shortage. And food shortages were "so common in human prehistory and history," write Brown and Konner, "that they could be considered a virtually inevitable fact of life in the past." Twice as many women survived the 1846 Donner party disaster as did men, it is worthwhile remembering—a gruesome statistic that derives from the different fat stores of men and women and, ultimately, from their different reproductive strategies.

And while periodic food shortages were probably the climate in which this sexual dimorphism in fat distribution evolved, this dimorphism is still important even in the places where food is plentiful and predictable and where the threat of obesity is far greater than that of starvation. Underweight women, or women who do not gain weight during pregnancy, are at risk of having babies with low birth weights, and low birth weight is associated with a whole host of

potential problems, including complications during delivery, physical and mental defects, and early death. Of those infants who die
before their first birthdays, about two-thirds are low-birth-weight
babies.

So if you've ever wondered why women seem to have a much
more complex relationship to food and eating than men—why
women seem to spend so much more time thinking about food and
why they are much more prone to eating disorders—the answer
must begin here in this distribution of soft tissues, here with the fact
that *all* females have a different relationship to food than males.

Seven

WHAT'S IN A MEAL?

While the sixty thousand murres on Coats Island were learning how to get along with their mates and neighbors on narrow ledges, the six humans on the island were adjusting to their cramped camp quarters. Our way was smoothed by the fact that each of us spent most of every day alone in a blind observing the birds and by heavy doses of camp humor having to do with the food, the filth, the number of mosquitoes that one was ingesting along with dinner, and the idiosyncrasies of the group: the fact that one of us, a fitness fanatic, ate so much because he was denying himself the fats that his lean body craved or that another never bathed or washed his clothes.

Nothing rots in the Arctic and there is very little chance of infection, so hygiene is not what it is in more temperate climes. "Even washing cups and pots was not a required practice," I once heard an Inuit, who had grown up in snow and sod houses, say. "We used unwashed cups when drinking tea at the neighbor's dwelling," he added. "We even shared chewing gums." At Coats, we never went that far, but the air in the tents and the cabin was definitely rich, and one didn't want to examine too closely the state of the countertops or the dish towels. My first night in camp, I went on a cleaning rampage, sweeping the cook tent, scouring the pots and pans, and soaking all the dish towels in boiling water. But the next day the towels smelled just the same, and there seemed to be just as many dead mosquitoes on the floor. Later I learned that Tony *always* put newcomers on kitchen duty in order to take advantage of their sense of cleanliness while it was still intact, before it had become diluted by the realities of camp life.

At breakfast and lunch, everyone cooked for themselves, but there was always a leg of caribou on the counter, courtesy of Josiah, the Inuit who lived in Coral Harbor. Members of the group ate it raw and blue, slightly cooked and red, well cooked and pink, or not at all according to their taste, background, and the state of the leg as it sat out day after day until it was finally turned into stew and dumplings by Tony about a week after Josiah first removed it from the ice chest in the snowbank. I was the "not at all" group, and I'm a little ashamed to admit that I was not more gastronomically adventurous in my travels to the far north, but the sudden change to a largely meat-based diet was too much for my vegetable-loving brain.

So at the same time that I was learning about the murres' diet and the murres' feeding behavior, I was also learning about my own. I tried to ignore the leg of caribou when I was in the cook tent, but there it sat, challenging my food preferences and my ideas about what constitutes a healthy diet. Like the health fanatic, who was trying to keep to his low-fat, muscle-building diet in this part of the world, where the consumption of adequate fat has always been essential for survival, there I was trying to hold on to my

ideas about the importance of an adequate intake of fresh vegetables. Fortunately, the plane I arrived on was carrying five hundred
pounds of resupplies, including many heads of cabbage and sacks
of carrots.

The baked-apple berries wouldn't be ripe until August, so I
couldn't supplement those vegetables with fresh fruit, but I did
find mountain sorrel as I explored the wet, rocky seeps near the
camp. Mineral- and vitamin-packed mountain sorrel is one of the
most important plants in the plant-poor Inuit diet, and it is also
sought after by geese, musk oxen, caribou, lemmings, voles, and
hares. Like these animals, I ate the small crisp leaves right there on
the spot, but the Inuit also preserve them in seal oil for the winter.
Once I was rewarded in my hillside climbing and foraging by the
sight of three fox kits scampering around the entrance to their den
at the bottom of a rock slide or scree. I knew it must be their den
because the area was littered with the bones and feathers of murres,
those unfortunate birds that had decided to try and expand the
boundaries of the colony rather than to wage a struggle for one of
the safer interior ledges.

At night in camp, the six of us took turns cooking, and what
each of us managed to produce on the tiny kerosene-fired stove
was a humorous study in contrasts. The health fanatic prepared a
low-fat shrimp loaf with canned shrimp. When it was my turn, I
cooked lots of rice, cabbage, and onions and just a small amount of
meat—one pork chop for each of us. When Josiah cooked, those
proportions were completely reversed. Each of us received one
very small piece of carrot and potato to accompany a very large
piece of roasted caribou. I wanted to ask for more vegetables to
accompany the meat that covered my plate, but a quick glance in
the pot told me that those were the only carrots and potatoes that
Josiah had made. I ate as much of the caribou as I could and hid

any dissatisfaction I felt, but it was many months, I have to admit—since this seems to be the time to confess to all of my dietary nearsightednesses—before I realized that Josiah had been every bit as unhappy with the meal that I had prepared as I had been with his. My "balanced" offering of vegetables and meat didn't look any more like a real meal to him than his did to me.

Nor would either of our meals have looked like meals, let's say, to a Japanese, who might have appreciated my heaping mound of rice but surely would have been disgusted by the large slab of meat and the huge helpings of vegetables. Or to a Brahman in southern India, who would have objected to both the pork and the caribou. Humans survive and thrive on such different foods in different parts of the world, on such different proportions of meat, starch, and vegetables, that it seems only logical that different people have different physiological needs and different digestive systems. It would make sense if vegetarians in India, whose diet can be ninety percent carbohydrate, have reduced body needs for protein or that the Inuit have special adaptations allowing them to live on diets that are mostly meat.

But research has shown that, despite the wide variation in human diets, humans exhibit very little variation in diet-related physiology. We all need the same forty or fifty different nutrients, and we can only survive on diets that fulfill all of those needs.

Many are familiar with the few examples of digestive variability that do exist—lactose intolerance, for example, the inability to produce lactase, the enzyme that breaks down sugars in milk, after weaning. Sixty-six to ninety percent of people of Asian and African descent are lactose intolerant, but only six to twenty-four percent of people of European descent, a reflection of the fact that there has been a long tradition of herding and milking in Europe and Eurasia, but not in the Far East and Africa. People in Europe have been dependent on milk and milk products for protein and essential vitamins for thousands of years.

Also familiar is celiac or wheat-eating disease, a sensitivity to the gluten of cereal grains found in people who have only recently been introduced to wheat as a dietary staple. Then there is the sucrose intolerance found in the Inuit population. The traditional Inuit diet, high in meat and low in carbohydrates, has been associated with the absence of the normal hormonally mediated insulin response to sugar consumption in three to ten percent of the population, resulting in a high degree of alcohol intolerance. In other human populations, researchers have also found variations in kidney function; in the levels of certain digestive enzymes; and in the length of the gut, for the intestines of people who eat a diet of mostly plant foods tend to be longer than those in heavy meat-eaters. But by and large, all humans have the same dietary equipment and the same dietary requirements, popular books on eating for your blood type notwithstanding. And these requirements have not changed much since the appearance of modern Homo sapiens more than forty thousand years ago.

We do, however, satisfy these requirements in entirely different ways. That is what is meant by dietary flexibility, a trait that Homo sapiens has in spades, a trait that has allowed us to populate (and overpopulate) every region of the earth. Our need for vitamin C is one example of this dietary flexibility. Most mammals do not need to eat foods containing this important vitamin (which performs essential work as a coenzyme and antioxidant and in collagen formation) because they synthesize vitamin C themselves, in their own bodies. Primates lost their ability to synthesize vitamin C, it is often argued, because they evolved in the tropical forest, where foods, fruits and leaves, are naturally high in this compound. Synthesizing vitamins is expensive, the argument goes, and those animals that don't have to synthesize vitamins, yet can live in perfect health, can put that energy into something else. The earliest forms of life were single-celled organisms capable of manufacturing all the compounds they needed for their survival. But, over time, life-forms evolved that were increasingly dependent on external sources of nutrients.

So how is it possible that humans, whose ancestors lost their ability to synthesize vitamin C because they lived in forests filled with fruit, can also survive in the Arctic, where there are no forests and very little fruit? The Arctic explorer Vilhjalmur Stefansson discovered how, after several Arctic and Antarctic expeditions foundered on that very question, losing all or most of their men to scurvy. In the 1930s, in a celebrated, medically supervised experiment, Stefansson and his associate Karsten Anderson lived healthily for an entire year on animal flesh alone. The experiment took place, not in the Arctic, but in the temperate regions of New York City. But Stefansson had the idea for it on one of his trips to the north, when he realized that the Inuit themselves were unaffected by the lack of fresh fruits and vegetables in their diet. The trick, he found, the trick he duplicated in Manhattan, was that they eat their meat raw or cooked so slightly as to not destroy the small amounts of vitamin C that are naturally found in meat but are very sensitive to heat. By the time humans began to inhabit the Arctic regions, fire had long since been discovered and was used to prepare many foods, but the people that managed to survive in this new, cold environment did so only by avoiding fire—at least with regard to cooking meat.

Vitamin C is not the only essential nutrient that presents a problem in the far north, for how do the Inuit find enough carbohydrates to eat? Some people mistake carbohydrates—the complex sugars of rice, wheat, potatoes, and fruits—for second-class, nonessential foods because they are low in protein or in complete protein. But protein isn't the beginning and the end of good nutrition, as it has taken nutritionists and the public a good long while to learn. Carbohydrates are necessary to provide the human body with a steady source of glucose, a principal form of energy for the human body and the *only* form that can be utilized by the brain and nervous system. Carbohydrates quite simply are fuel for the brain, and a rapid drop in the continuous supply of glucose to the brain brings about confusion, coma, and ultimately death. The human brain alone needs 100 to 145 grams of carbohydrate every day, the amount of carbohydrate in about seven slices of bread. Now fats can be used as energy for the body's muscles, but they can't be used to produce glucose without disturbing the body's normal acid-base balance, a condition that occurs during fasting. And glucose can be synthesized from protein through the breakdown of amino acids into their glucose and amine-containing parts, but an all-meat diet typically provides only about ten grams of glucose for every 2,500 calories ingested.

To meet their glucose needs, the Inuit would have to eat huge amounts of meat. But eating huge amounts of meat can itself be a problem, especially if the animals being eaten are very low in fat, as is the case in inland parts of the Arctic in winter and early spring, when the fat reserves on caribou and rabbits are extremely low. This is a problem that Stefansson was also aware of. "If you are transferred suddenly from a diet normal in fat to one consisting wholly of rabbit, you eat bigger and bigger meals for the first few days until at the end of about a week you are eating in pounds three or four times as much as you were at the beginning of the week," Stefansson wrote in his *Arctic Manual.* "By that time you are showing both signs of starvation and of protein poisoning. You eat numerous meals; you feel hungry at the end of each; you are in discomfort through distention of the stomach with much food and you begin to feel a vague restlessness. Diarrhea will start in from a week to 10 days and will not be relieved unless you secure fat. Death will result after several weeks."

Consumption of large quantities of low-fat meat leads to a condition called "rabbit starvation," a condition described by Arctic and United States explorers alike. "We tried the meat of horse, colt, and mules, all of which were in a starved condition, and of course not very tender, juicy, or nutritious," Randolph B. Marcy said of his travels across the prairie lands of North America during the winter of 1857–58. "We consumed the enormous amount of from five to six pounds of this meat per man daily, but continued to grow weak and thin, until, at the expiration of twelve days, we were able to perform but little labor, and were continually craving for fat meat."

No, protein is not the beginning and end of good nutrition. Adequate protein is absolutely essential for health and well-being, and protein, as researchers have recently found, is more satiating than either carbohydrates or fats, so the person who eats healthy amounts of it can spend more time doing things other than eating. But too much protein quickly becomes a digestive nightmare. Most human populations today obtain between ten and twenty percent of their total energy requirements from protein, says John Speth, an anthropologist at the University of Michigan. And there seems to be a safe upper limit to protein consumption of about fifty percent of total calories. In the United States, where many people can eat as much meat and protein as they like, most seem to naturally limit their protein intake to less than twenty percent of their daily calories.

Part of the problem with protein has to do with the rate at which the liver can metabolize amino acids and the body can excrete excess nitrogen (large amounts of protein overload both the liver and the kidneys), and part with the rise in the body's heat or metabolism as it consumes protein. This rise in metabolism, as Speth and anthropologist Katherine Spielman explain, can be as high as thirty percent, versus about six percent for fats or carbohydrates. Which means that for every one hundred calories of protein eaten, thirty more are needed just to compensate for that metabolic increase. This is the reason why high-protein diets seem to work, though in fact they burn more muscle than fat.

When Stefansson's New York diet was lean, raw meat with only three percent fat, he felt hungry and ill and his metabolic rate was far above normal. He improved his health by consuming increasing amounts of fat and was the healthiest when three-quarters of his calories came from fat and only one-quarter from lean meat. Personally, I cannot imagine eating that much fat in New York City, especially in the dog days of August, but by doing so Stefansson discovered how the Inuit avoid both scurvy and rabbit starvation. They consume enough fat to keep their metabolisms low and to provide their bodies with energy so that the protein they do eat is spared for the jobs of body maintenance and supplying glucose to the brain. Carbohydrates would do the job just as well, or even better. Carbohydrates are the most efficient way of supplying the brain with adequate glucose, but the Inuit have rarely had that option. Given the unforgiving nutritional constraints of humans, though—that the brain runs on glucose and that there is a limit to the total amount of protein (plant and animal combined) that a human can safely consume on a regular basis—it's not surprising that inland communities in the Arctic have always had an active trade with coastal communities. They exchange caribou meat and skins for seaweed, as already mentioned, and for the blubber of seals and whales.

Strict carnivores, by the way, handle this same problem by having smaller brains that don't require as much glucose and with

specialized enzymes that improve their ability to metabolize protein. But they too show signs of craving fats during certain times
of the year. Much like the Indians of the Southern High Plains,
who killed bison in the spring and ate just their fatty tongues,
polar bears often strip seals of their blubber, then leave the carcass
for other animals to eat. Leopard seals in Antarctica strip Adélie
penguins of their fatty skins. Usually, these behaviors are attributed to the fact that fat contains almost twice as many calories as
the same quantity of carbohydrate or protein, which is true, but in
many situations, calories are not nearly as important as the realities
of protein metabolism.

So the Inuit are able to survive in the far north on a high-
protein diet because they eat their meat raw, or something very
close to it. And because they eat large quantities of meat and fat
and supplement their diet with seaweed and whatever plants and
berries they can find, including those in the stomachs of the herbivores they kill. Elsewhere, humans have managed to survive by
adopting elaborate food-processing techniques—another form of
dietary flexibility. This is especially true in areas where people
depend on agricultural crops for most of their calorie needs. There,
shortages of meat in the diet can cause protein-energy malnutrition, an often fatal condition.

Humans need a little less than a gram of protein for every
kilogram of body weight every day (that is about forty-eight grams
of protein, the amount of protein in a cup of roast chicken, for a
woman weighing 130 pounds). Children, pregnant women, and
adults who are ill need more. If we don't get it, we turn to our
own tissues to provide the amino acids necessary for making all the
different kinds of proteins that humans need to digest, think, fight
infections, and so on. The human body cannot store protein for
any length of time, so any excess eaten one day is broken down and

used to produce energy. Sources of protein are not all equal, as we know. The highest-quality proteins are those that provide all of the essential amino acids (the amino acids that we cannot synthesize) in the same proportions that they are found in the proteins that make up our bodies. The highest-quality proteins are animal proteins found in eggs, milk, and meat. Proteins derived from plant sources are usually deficient in one or more essential amino acids.

The Indians of America and Central America, as many are aware, found a way around the deficiencies of individual plants by combining different foods in their plant-based diet. Corn is deficient in the essential amino acid lysine, and beans are deficient in methionine, but if corn and beans are eaten together, each compensates for the other, and the fortunate diner gets a full complement of amino acids. Once populations of wild animals in America and Central America were severely reduced by hunting and overhunting, discovering this combination was the key to survival in this part of the world. Or so many of us believe. In truth, mutual supplementation, as it is called, wouldn't by itself be enough to sustain human populations. A corn and bean diet (even one enriched by squash and other plants and fruits) is usually inadequate in vitamin B3, or nicotinic acid, which the body synthesizes from the essential amino acid tryptophan. In order to survive and thrive on their diet, the Indians of Mexico and Central America also needed to add a culinary twist.

Corn actually contains enough tryptophan to sustain human populations and prevent them from developing pellagra, another often-fatal dietary disease, but the tryptophan in corn is normally bound to the indigestible portion of corn. It normally passes right through the body and is inaccessible to the body's digestive enzymes. But if corn is treated with an alkali before it is cooked and consumed—if, for example, it is boiled with lime (an alkaline earth containing calcium oxide)—the tryptophan is released from the indigestible portion, and the quality of the protein fraction of corn is significantly improved.

"Why do you boil your corn with lime?" Solomon Katz, a professor of pediatrics and anthropology at the University of Pennsylvania who has written extensively on the evolution of cuisine, asked a number of Mexican women. Katz collected data on corn consumption in numerous societies and found a striking correlation between the degree of a society's dependence on maize and its use of alkali-processing techniques. But the answers he got to his question were that tortillas just didn't taste right if the corn hadn't been cooked with lime . . . that they were more difficult to roll out . . . that the hulls of the corn kernels were more difficult to remove. The women seemed to be unaware of the nutritional benefits of their methods of cooking corn. No one knows how these methods came about, but let's say a village just happened to make its pots out of limestone or to boil its corn in water that contained a lot of lime. The people in this village would be better able to survive those periods when game was difficult to find and they had no animal source for the B vitamins. Before long, one might expect that everyone in the area would have a taste for corn treated in lime.

Unfortunately, when corn was introduced into Europe in the seventeenth and eighteenth centuries by explorers who were impressed with its high yields in the Americas, its promoters failed to also introduce the technology that makes corn a nutritionally sound food. As cultivation of corn spread among the poor of Italy and France (and much later, among the sharecroppers of America), so did pellagra. People in Italy used to say that pellagra gives rise to seven kinds of ills: "It drives one crazy, it drives one into the water, it draws one backwards, it makes one walk bent, it gives one vertigo, it gives one ravenous hunger and it causes rashes on the skin." They left out the final wasting stage, in which the disappearance of all subcutaneous fat leads to diminished strength, diarrhea, and death from weakened heart muscles or tuberculosis, an infection to which pellagrins are particularly prone. As recently as 1930, there were some 200,000 people in this country alone suffering from this easily remedied disease.

Perhaps the most remarkable example of human dietary flexibility and inventiveness is the dependence of many Asian populations on the soybean, an annual legume that, because of its nitrogen-fixing properties, had been used to improve the soil long before it was used for food. Soybeans, it would seem, are the healthiest of foods; but as Western populations discovered when they tried to divorce soybeans from the culture in which this food had evolved, unprocessed soybeans—soybeans straight out of the ground—are inedible.

At the end of the Second World War, as part of the Marshall Plan, the United States shipped large quantities of soybeans to the hungry populations of Korea, Japan, and Germany. The Koreans and the Japanese thrived on these donations of food, but the Germans became sick whenever they ate the beans, suffering stomachaches, weight loss, and other ill effects. Health scientists concluded that there must be a physiological difference between these populations that accounted for their difference in ability to digest soybeans, and the Americans stopped their shipments to the Germans. But the difference, it turned out, was cultural rather than biological, cooking rather than physiology. As food scientists in the United States tried to understand what made soybeans a health food in some countries and a bellyache in others, they found that soybeans contain a potent compound that blocks the breakdown of proteins. This compound, or antitrypsin factor, was what was making the

Germans sick. It could be deactivated at high temperatures or by prolonged boiling, but high temperatures and prolonged boiling also destroyed most of the food value of the soybeans—and the reason, therefore, for the food shipments.

Then these food scientists made a major breakthrough, or so they thought. They discovered that this antitrypsin compound could also be removed by a brief period of boiling, followed by precipitation with magnesium or calcium salts. This protocol reduced the cooking time by two-thirds, and what was left of the beans, the precipitated curd, was a completely digestible, high-quality protein that, if consumed with rice, provided a well-balanced diet. With great excitement, the scientists presented their findings to the world, only to discover that the Koreans, the Japanese, and the Chinese had been using the same technique for many centuries in their preparation of tofu, or soybean curd.

And it is this technique, plus the fermentation of soybeans to make soy sauce, that has allowed Asian societies to maintain high population densities even in the absence of large amounts of available animal protein. Animal products are the source of the highest-quality proteins, but of equal importance is the fact that they are the only source, the only naturally occurring source, of another essential vitamin—B12. Without adequate amounts of B12, humans come down with pernicious anemia, which is associated with spinal cord lesions, weakness, diarrhea, and a numbness in the arms and legs. But fermented foods—soy sauce, beer, yeast—also contain vitamin B12, and since the human body needs only a very small amount of vitamin B12 daily, the regular use of soy sauce as a condiment is more than adequate to meet those needs. Our dependence on B12 and its limited availability (almost every other essential nutrient can be found in plant foods) also tells us something important about the origins of the human diet. Our original diet had to contain small, daily amounts of animal products. Either that or soy sauce or beer. Pure vegetarianism is very rare in our species and only as recent as the continuous availability of staple plant foods. It is dependent on that continuous supply and on learning, by trial and error, and over many generations, how to make plant-based diets suitable for human nutritional needs.

So yes, we all have the same dietary requirements and pretty much the same dietary equipment. What differs is how different peoples in different parts of the world go about meeting those requirements. Cultural adaptations like precipitating soybeans with magnesium, boiling corn with lime, or eating meat on the rare side are the human animal's forte.

There are plenty of other animals with elaborate food-finding techniques. South American bola spiders attract moths to their snares by producing chemicals similar to the pheromones used by female moths to attract their mates; the leaf-cutter ant of Central and South America harvests leaves, not to eat itself, but to bring back to its nest to feed to the fungus that the ants do eat. But no other animal has adaptations so numerous and so diverse. No other animal's feeding adaptations allow it to inhabit almost every region of the earth. The course of cultural adaptation and evolution may be driven just as much by accident and chance as the course of natural selection. But because cultural changes are not attendant on the happenstance of random mutations in the genes, they take place much faster than biological changes. And these changes can be kept alive, not by the replication of DNA molecules, but by learning, memory, and language, more workings of our very large brains.

Eight

THE WORM AND THE APPLE

The intelligence that allows humans to feed whole societies on the formerly indigestible soybean did not spring up de novo with human beings. We may have the largest brain per body size of any animal, but large brains are a trait shared by the order of primates—as are complex feeding strategies. The toque monkey of Sri Lanka feeds on caterpillars that are abundant in adina trees but that escape most predators by dropping down on a silk thread as soon as they sense movement nearby. The toques, however, have learned to simply fish the caterpillars up by pulling on this same thread, pulling faster than the caterpillar is able to descend. Chimpanzees in different parts of Africa eat very different foods, and they sometimes eat the same foods in very different ways, behaviors that many primatologists have argued are incipient forms of the same kinds of cultural adaptations at which humans excel. Those primates that most closely resemble the ancestral primate stock don't show much behavioral flexibility—

the nocturnal mouse lemur, for example, can't unlearn associations that it has made—but flexibility is the stock and trade of the later-evolved monkeys and apes.

Fifty-five million years ago, when flowering trees were first beginning to spread across the planet, some small mammal in Africa climbed up into one of those trees in search, perhaps, of insects. But over time the descendants of this mammal came to rely more and more on the edible parts of the trees themselves, a new arboreal niche that had not yet been filled. Over time they radiated out to control a large proportion of the highest-quality foods in the tropical forest—fruits, flowers, and young leaves. And they developed the traits characteristic of primates: hands that are well suited for grasping slender branches and manipulating shoots and other delicacies; visual acuity, depth perception, and color vision that enable them to move quickly through the three-dimensional space of the forest canopy, discerning the presence of any ripe fruit or tiny new leaves; brains that are much larger than expected for a similar-sized terrestrial mammal and that allow them to process all of this information and learn the location of trees with edible parts, their flowering and fruiting times. Entry into this new dietary niche appears to have placed considerable pressure on animals to lower the costs associated with finding foods. And one way to do this was to increase cerebral complexity.

It used to be thought that finding food in the tropical canopy was as simple as reaching out a hand and taking it. It used to be thought that monkeys and apes led something of the *Life of Riley*. But slowly it has dawned on primatologists and anthropologists, as they have actually followed these animals around in their daily quest for food, that finding enough to eat in this environment is anything but simple and that primates may have their large brains for reasons associated with this quest.

As a graduate student in anthropology in 1974, Katharine Milton, a native of Alabama who speaks with a distinct southern twang, visited a tropical forest on Barro Colorado Island in the Republic of Panama. There she expected to observe monkeys in all their indolent splendor. Yet during her first weeks of trailing behind mantled howler monkeys—named for the mantle of chocolate-brown fur that drapes their dark shoulders and for their loud, unearthly howls—she realized that they were not behaving as expected. "Instead of sitting in a tree and eating whatever happened to be growing nearby," Milton later wrote in an article on diet and primate evolution in *Scientific American*, "they went out of their way to seek specific foods, meanwhile rejecting any number of seemingly promising candidates. Having found a preferred food, they did not sate themselves. Instead they seemed driven to obtain a mixture of leaves and fruits, drawn from many plant species."

The old, easy-living dogma about primate life was far too simplistic, Milton realized, and on the spot, she decided to learn more about the problems facing howlers as they went about meeting their nutritional needs. She had found her dissertation project and would spend the next twenty years pursuing monkeys in the Panama forests, preparing herself for an insight that would become a bold new view of the purpose of brains and their role in human evolution.

Lush and green as the tropical forest is, Milton, now a professor of anthropology at the University of California at Berkeley, came to understand that it presents tree-living primates with a number of nutritional dilemmas. Unlike grasses or low-growing bushes (the foods on which most terrestrial herbivores feed), trees are widely separated, especially trees in fruit or flower. Primates cannot just leap from branch to branch to feed. Rather, they must have a mental map of their territory, an awareness of when those trees will be producing flowers or fruit and for how long.

Even those primates that eat a diet primarily of leaves have to be selective about just which leaves they eat, for leaves can be full of toxic compounds that can make an animal ill or interfere with its digestion. Many of us are familiar with the fact that rhubarb leaves contain oxalic acid or that peach pits and bitter almonds contain cyanide. These are just two of the many chemical compounds that plants have evolved in order to protect themselves from being eaten (the antitrypsin factor in soybeans is another). At best these chemicals taste awful (but can have medicinal effects or can be highly addictive, like the nicotine in tobacco or the cocaine in coca leaves); at worst they are lethal. Primates must learn either to avoid plants laced with toxins or to eat them in small enough quantities that enzymes in their liver or bacteria in their guts can detoxify the poisons. They must learn to eat leaves when they are young and before they are fully loaded with these protective devices. (Terrestrial herbivores, by the way—cattle, deer, moose, and sheep—do not need to be as concerned with avoiding toxic compounds since the bacteria and other microorganisms in their complex compartmentalized stomachs can detoxify most compounds. These herbivores live entirely on the waste products of their bacteria, but it is still to their advantage to feed those bacteria well and not to tax them with too many poisons. So terrestrial herbivores, too, tend to avoid plants that are loaded with toxins or to eat them only in small quantities.)

A diet of leaves and other plant parts presents another problem for primates, and that is fiber. Plant cells are encased by walls made up of cellulose, and cellulose is resistant to mammalian (but not bacterial) digestive enzymes. In the United States, we have been told to add fiber to our diet because it slows down energy intake and helps to regulate bowel activity and remove unwanted compounds from the body. But for the primate in the wild, the primate that has a hard time finding energy enough to fuel its search for food and all its other social and reproductive activities, fiber can be a big problem. It takes up room in the gut, yet it provides little energy. Mature leaves are sixty percent indigestible fiber, which is another reason why many primates prefer leaves when they are young and tender.

Primates also have trouble obtaining a complete balanced diet. Fruits are full of easily digested carbohydrates, but they have very little protein. Flowers and leaves have protein and essential fatty acids but few carbohydrates. And plants, unlike the animal prey eaten by the terrestrial mammals from which primates evolved, rarely contain the full complement of amino acids that most mammals require in their diet. In order to obtain all the essential amino acids, primates need to eat from complementary protein sources within a fairly short time of each other and to supplement those plant foods with insect and animal prey. The French biologist Claude-Marcel Hladik found that chimpanzees on an island in Gabon eat the leaves from two particular trees throughout the year and that the amino acids in those leaves are at least partially complementary. Together they are still low in some amino acids, but those are provided by the ants that the chimps also eat all year long. (Terrestrial herbivores deal with the problem of complete protein in much the same way that they deal with toxic compounds: the bacteria in their guts are able to synthesize all the amino acids. But again, it is to an herbivore's advantage to supply those bacteria with a healthy mix of nutrients, nutrients that best promote bacterial growth and function and that are rich in amino acids.)

To eat right in the tropical forest, in other words, those animals that took to the trees had to become omnivores. They had to eat a variety of different foods, some from plants and some from animals. Primates are primarily herbivorous, but most supplement their diet of greens and fruit with insects and animal prey. Monkeys in tropical forests had often puzzled biologists by their frequent habit of taking just a bite or two out of a piece of fruit before tossing it to the ground. Sometimes they have been observed sending down a near-constant rain of half-chewed pieces of leaves and fruit, a behavior that has contributed greatly to their *Life of Riley* image. Recent studies, though, have shown that these monkeys are not being wasteful in their eating habits but rather extremely finicky. What they were after was not the fruit, or not just the fruit, but the insects eating the fruit—the worm in the apple.

Or the worm *and* the apple. In seeking out these insect-containing fruits, monkeys also resemble humans when they combine energy-rich carbohydrates and protein-rich meat to make, say, a sandwich or a tamale, stuffed dumplings, or a stew of meat and potatoes, or when they carefully alternate bites of meat with bites of tuber or plantain, as hunter-gatherers have been observed to do. All of these behaviors spare proteins for body building and body maintenance and keep them from being broken down for energy. Chimpanzees, too, manifest this protein-sparing behavior when they roll or wad up eggs, fledglings, or meat with leaves and bark, then chew it all together. Fatty meat would do away with the necessity for chimpanzees to make roll-ups of their food, for hunter-gatherers to take alternate bites, but the meat of most wild animals, as already mentioned, is very lean.

An exception to the rule that primates eat from many different foodstuffs, and one that reveals a great deal about the course and trade-offs of primate evolution, is the colobus monkey, a long-tailed monkey of Africa and Asia. These monkeys eat very little but leaves, living the way people used to think all monkeys and apes lived. But colobus monkeys are able to survive on leaves only because they have a stomach like that of a cow. Their stomach is not at all like the simple acid stomachs of most other primates. Instead, it is compartmentalized into two relatively separate structures, one of which is maintained at a slightly basic pH so that bacteria can ferment the fiber in leaves and turn it into the fatty acids that the colobuses use for energy. So not all primates need to eat nutritionally balanced diets; not all need to avoid plant toxins or keep track of trees and their flowering and fruiting seasons. The colobi are perfectly adapted to their slow, ruminating life, but such perfect adaptation can have its drawbacks. The colobi must have lots of leaves to feed to their bacteria. If their environment should dramatically change, they could not survive on any other kind of food.

And this gets back to Katharine Milton and the insights she gained into primate life and primate evolution while following monkeys around the Panama forest.

Milton began by observing just howler monkeys, and she spent three years tracking a group of them as they searched for leaves and fruit from as many as twenty-five different species of plants a day. She was up with them in the morning when they first began to stir at 4:30, and she tailed them until six in the evening when they settled down for the night. Some of their foods, like the newly emerging tips of the leaves of the enormous *Ceiba pentandra* tree, were edible for only a few hours a year; others were available for longer periods. Unerringly, the howlers tracked them all down.

Despite their diverse diet and their unfailing sense of direction, though, and despite their howls, howler monkeys are somewhat placid and dull. And as Milton pursued them, she couldn't help becoming intrigued by a second resident of the Panamanian forest: the spider monkey. These speedy animals used to race past her like greased lightning as she sat under a tree where her sedate howlers were either eating or resting, their bodies and tails draped languidly around the branches. Spider monkeys are about the same size as howlers, fifteen to eighteen pounds, and they are closely related to howlers, but their temperament couldn't be more different. Howlers travel through the forest canopy on all fours; spider monkeys swing from the branches like Tarzan. Howlers seemed very uninterested in Milton or in the peanut butter sandwich that she brought with her for lunch each day; spider monkeys are playful and mischievous and had no trouble recognizing Milton's lunch as food. "They'd swing down toward you; they'd threaten you," Milton once told a reporter. "They have a keen idea of what a peanut butter sandwich is. You simply cannot eat in front of them."

Milton decided to add spider monkeys to her observations, though keeping up with them was a great challenge. Their territories were at least ten times the size of those of howler monkeys, 750 acres versus seventy-five, and every day they would comb them for the hundreds of different kinds of fruit that make up their diet. Why are spider monkeys and howlers so dissimilar? Milton wondered. Does it have anything to do with the differences in their diet? Most of the year, howlers divide their feeding time about equally between fruit and leaves, but when fruit is hard to find, they eat nothing but leaves. So how, she mused, do howlers get enough energy to live even their placid lives? And why do they sometimes ignore fruit even when it is readily available? How do spider monkeys, on the other hand, get enough protein from their diet of mostly fruit?

After analyzing the digestive tracts of monkeys that had died in the forest and discovering that the colons of howlers were considerably wider and longer than those of spider monkeys, Milton was able to answer most of these questions. Food has to travel much farther and remain much longer in howler guts, which means that their digestion is slow enough that bacteria have a chance to ferment the masses of fibrous leaves. Howlers don't have anything like the compartmentalized stomachs of colobus monkeys, which exist almost entirely on leaves, but more than thirty percent of their calories come from the energy-rich fatty acids that bacteria in their guts produce. The digestive system of spider monkeys, in sharp contrast, is as speedy as those monkeys themselves. Transit time is twenty hours in howlers but a quick four hours in spider monkeys. This doesn't allow the spider monkeys to turn fiber into energy, dross into silk, but they don't need to. They get all of the energy they need, and a little of the protein, from fruit. The tender young leaves, with which spider monkeys supplement their fruit diet, provide the rest.

Each monkey's diet, Milton realized, perfectly matches its digestive physiology, its lifestyle (a howler monkey's diet couldn't provide the fuel that is necessary to propel a spider monkey around)—and its intelligence, she kept thinking as she had to work harder and harder to outsmart spider monkeys and keep her lunch for herself.

Curious as to whether there was a difference in the brain sizes of these two species, Milton consulted the statistics of physiologist Daniel Quirling and found that spider monkey brains weigh twice as much as howler brains, 107 grams versus 50.4. And while brain size isn't a wholly accurate measure of intelligence (neuronal complexity and density are also important factors), the magnitude of this difference was too large to be ignored or explained away by all the factors that affect brain function. The news didn't really come as a surprise to Milton, and yet, she says, "It was a eureka moment. Here were two monkeys, the same size, living in the same forest, but so different. Compared with the howlers, spider monkeys were brighter and more lively. They matured more slowly and had more to learn; they made more ruckus, with a greater variety of vocalizations; they ate widely dis-persed, high-energy foods that were harder to find—and their brains were twice as large." The solution to the problems of diet seems to have made these animals what they are.

Later in her *Scientific American* paper, Milton proposed that there seem to be two basic strategies for coping with the problems of a tree-living, plant-eating life. In one, the strategy favored by colobus monkeys and, to a lesser extent, howlers, natural selection has favored the acquisition of anatomical specializations, especially of the digestive tract. Specializations like a compartmentalized stomach or an enlarged colon allow primates to depend on the most readily available plant parts: mature leaves. In the second strategy, selection has concentrated on changes in an animal's behavior—and in its intelligence—so that high-quality foods are easier for it to find. (A third possible strategy, eating mature leaves *and* growing a bigger body and brain, has not been adopted by any primates probably because of the limits placed on size by arboreal life. Some terrestrial and sea animals, notably elephants and por-poises, have used this strategy though, and it works because of a curious, imperfectly understood nutritional law. Big animals must consume greater amounts of food than small animals to run their big bodies, but their metabolic costs per unit of body weight are less than those of smaller animals. Most large animals have used these energy benefits to move into a new dietary niche. Because of the energy savings associated with size, large animals can survive better on lower-quality foods than small animals. But some have applied those energy savings toward a larger brain.)

As a group, primates tend to depend on wit rather than on witless microorganisms to cope with life in the trees (and on high-quality foods rather than size). *All* primates have larger brains than

their similarly sized terrestrial counterparts, but a comparison by
Milton of brain sizes in the order as a whole has found that those
species that eat higher-quality, widely dispersed foods have larger
brains than those that feed on lower-quality, more uniformly distributed resources. It is fashionable these days not to compare
intelligence in animals, to take the point of view that each animal
has the brain it needs to survive in the environment in which
it lives. But this point of view belies the importance of diet in
shaping intelligence—and belies consideration of those turning
moments in evolution when a change in diet might have precipitated a surge in cerebral complexity.*

As Milton envisions this process in spider monkeys: "It would
have been a feedback process in which some slight change in the
monkeys' foraging behavior conferred a benefit, which in turn per-
mitted a modest improvement in the quality of their diet, which led
to an excess of energy. Over generations, the monkeys that spent the

---

*Or a decline in cerebral complexity, as the case may be. In an article in
Natural History, University of Idaho zoologist John A. Byers describes his
first look at a koala brain, seen during a sabbatical year in Australia. "As the
veterinarian lifted off the top of the koala's skull," he writes, "I was amazed
to see that the brain did not fill the space inside. The smooth cerebral hemi-
spheres—each about as large and thick as the peel from a quarter of an
orange—were so small that they did not meet at the midline. . . .

"At one point in the species' evolution," Byers explains, "the koala brain
undoubtedly filled its skull, and the current mismatch suggests that the
reduction in brain size was rapid." But why did it happen? The answer,
most probably, has to do with diet. "Koalas eat little other than nutrient-
poor eucalyptus leaves, and they have the low metabolic rate and slothlike
habits appropriate for animals with a leaf-eating lifestyle," Byers continues.
"Very few fossil koalas exist, and next to nothing is known about how
these creatures came to be strict leaf eaters, but biologists agree that their
ancestors must have had a more diverse diet. Reduction of the brain—a
metabolically expensive organ to maintain—may have been part of the
koala's adaptation to a low-energy diet."

energy on making their brain slightly bigger and more complex had an evolutionary advantage. Their improved brain allowed for more helpful changes in their behavior, and so on."

Could it be just an accident that primates travel these two different roads: the road of the gut and the road of the brain? Are there no primates that have large, slow stomachs *and* large brains? The answer is no, and Leslie Aiello and Peter Wheeler at the University College in London and John Moores University in Liverpool think they have the reason why. Brains are expensive organs to run. The brain of a human adult represents only two percent of an adult's body weight, yet it consumes eighteen percent of its energy. And brains must continually be supplied with energy in the form of glucose and oxygen, a task made more difficult by the brain's inability to store significant amounts of energy and by the fact that the brain is never at rest. So bigger brains cannot be had by just any leaf-eating animal wishing to contemplate its navel. Brains require a reliable energy source in order to grow bigger. There are tight tautologies here, important loops. Brains require a constant source of energy, but they also make finding high-energy foods much easier. Life in the treetops may have made brains necessary, but the high-energy, high-carbohydrate foods of the forest—nuts, fruits, and honey—also made them possible.

Brains, though, are not the only organs that are expensive to run, as Aiello and Wheeler explain in a paper published in 1995 in *Current Anthropology*, "The Expensive Tissue Hypothesis." Brains have very specific energy requirements, but all organs—guts, hearts, kidneys, and so on—are expensive to run. What links the size of guts and the size of brains in an inverse fashion in primates is the simple fact that animals eating high-quality foods can make do with much smaller, less-expensive guts than animals eating low-quality foods. All animals have to accommodate the costs of their various tissues within their total energy budgets. If an animal reduces the size of its gut, it may be able to apply that savings toward a brain.

Theoretically, this is not the only way that large brains could be financed. Animals could reduce the size of other organs, or if sufficient food were available, they could increase both their body and brain size or increase their metabolism to produce more energy. In weasels, for example, it is possible that at least part of the cost of their larger-than-average brain is reflected in their higher-than-average metabolic rate. A reduction in gut size is not the only way to balance the high energy requirements of a large brain, as Aiello and Wheeler readily acknowledge, but it is the most probable means by which primates did this.

Now is as good a time as any to point out that while primates, in general, have large brains, humans have brains that are about two times larger than expected even for a primate our size. And that the size of the human gastrointestinal tract is smaller than expected by about sixty percent (though there is some variation in human gut size and length depending on diet). "The increase in mass of the human brain," Aiello and Wheeler note, "appears to be balanced by an almost identical reduction in the size of the gastro-intestinal tract." The cost of our brains, in other words, and the energy savings of our reduced guts are almost the same.

Which leads these two English scientists to a simple conclusion about the forces that have shaped human evolution. No matter what was selecting for large brains in humans, whether it was the demands of group living, as some have suggested, or the very long time it takes for human children to mature into self-sufficient adults, or as Milton proposes, the difficulties of the food quest itself, large brains could not be achieved without a shift, sometime during the course of human existence, to a diet containing foods of a higher nutritional value. A higher-quality diet is what relaxed, in primates, and eventually humans, the metabolic constraints against brain expansion.

What is essential, in other words, in understanding how humans came to have such large brains is identifying the factors that allowed them to have such small guts. For spider monkeys, it was the simple, easily digested carbohydrates in fruit. For our ancestors, five to ten million years ago, confronting the changing environmental conditions of the Pliocene Era when forests were drying up but grasslands were expanding and, with them, populations of large herbivores,* it was incorporating increasing amounts of meat and fatty bone marrow into their diet *and* the discovery that the fleshy, carbohydrate-rich roots, bulbs, and tubers of plants are good to eat. Eating the flesh of animals much larger than ourselves is one way that human diets differ from those of chimpanzees, our nearest primate relatives, but consumption of these buried storage organs, which are in fact plentiful in the African savanna grasslands, is another. And while there is a tendency to put a great deal of emphasis on the role of hunting in shaping human existence and human evolution, the amount of protein that humans can safely eat (in the absence of enormous quantities of fat) is limited. Plus, meat, as we know, while an excellent source of amino acids, fats, and some vita-

---

*In his excellent book *The Origins of Virtue*, Matt Ridley explains why grass was so important to the emergence of large herbivore species. Grass, which first appeared about 25 million years ago, grows from its base, not its tip, so it is not easily killed by grazing. It is, in fact, helped—or fertilized— by grazing, so grass encourages rather than discourages animals from feeding on it. It doesn't defend itself with toxic chemicals or spines. But since it is a low-quality forage, it better supports large animals than small animals.

mins and minerals, is not a good source of energy for brains. There is a reason why strict carnivores don't have high IQs.

No one knows how this hidden, underground food source was first discovered by our human ancestors, but Mark and Delia Owens provide a possible clue in their description of the Kalahari Desert during a drought. There they saw African antelope pawing deep holes in the ground in search of fleshy, succulent tubers from which they could get enough moisture and nourishment to stay alive. Could this sight, one wonders, also have been witnessed back in the Pliocene by certain large-brained hominids? Unlike meat, some starches are not readily digested in an uncooked form, so cooking may have become important in hominid life at the same time as roots and tubers. The earliest secure evidence of the use of fire dates back 750,000 years, long before the emergence of Homo sapiens. But the purpose of those fires, whether they were for warmth, defense, or cuisine, is not known.

"You are what you eat," the saying goes, usually referring to an individual's diet or the diet of a particular culture. But we now know that it refers to entire species as well. "The food we eat," observes Milton, "makes us human." Our large and sophisticated brain, that which distinguishes us from other animals, is predicated on a diet of energy-rich, high-quality foods.

This conclusion of Milton and others tends to be substantiated by the fossil record. Australopithecus, one of the first genus in our family, were bipedal primates, but their brains were not much larger than those of apes today, *and* they had massive jaws and large molar teeth well suited for a diet of tough plant material. Homo habilis was similar in size to Australopithecus, but with a larger brain and smaller teeth, indicating that they were selecting easier-to-chew, less fibrous foods than Australopithecus and/or they were somehow processing foods to make them easier to chew and digest. Habilis was replaced by Homo erectus, with an even larger brain and even smaller jaw and teeth, and erectus by Homo sapiens.

I try explaining this coevolution of guts and brains to my fifteen-year-old daughter, who thinks that she can exist on a diet of salad and plain pasta—but with very little result. She would say that I am exaggerating about her diet, and of course I must be since she is growing and happy and doing well in school. Perhaps I harp on her about food because the cook in me doesn't like to see the meals I prepare go unappreciated. Or is it the science writer in me that doesn't want her going out into the world with the illusion that all foods are the same?

But as the French anthropologist Claude Lévi-Strauss has said, foods must be not only good to eat but also good to think, and the pot of stew or the leg of lamb that I have cooked is not good for my daughter to think. Human brain expansion may have been made possible by increased consumption of meat and carbohydrate-rich roots, but humans now have the luxury of letting our ideas and our beliefs shape our meals. We still have to meet all of our physiological needs, but we can do it in many different ways, picking and choosing among the many cuisines of the world, among the many different ways that people have solved the puzzle of these requirements. I don't mean to say that all diets are equally healthy. Some are lacking in important nutrients and cause deficiency diseases. Others, including the diets of many Americans, supply too much energy and cause diseases of overconsumption. But all human diets, even the rice- and cereal-based diets of people in South Asia and China, are much higher in protein quality and energy than the diets of other primates. In agricultural societies, people learned how to create complete, high-quality protein from vegetable sources. They need just small amounts of meat or milk to balance out their nutritional picture.

Yes, one can pick and choose from the diets of many cultures, but one has to be careful to consume the whole dietary package. I was reminded of this the other day when I was eating lunch in an Indian restaurant with my daughter and two of her friends. One of the friends had recently become a vegan and was avoiding all animal products (even refined sugars, since some manufacturers use bones to filter their product). In the restaurant, she helped herself to the vegetable curries and stewed lentils but refused the dishes made with yogurt and cheese. I was glad to see this young girl eating so heartily and so adventurously, since my daughter and her other friend had ordered nothing but salad and bread. And I knew I probably shouldn't raise the topic of ghee or clarified butter, the preferred cooking fat in India and the source for many Indians, along with cheese and yogurt, of vitamin B12. But I was curious. What did she think made Indian food so filling and delicious? The success of the human species is based on dietary flexibility, on our ability to find and add new, high-quality foods to our diet when old foods become scarce. But just how flexible are we as individuals?

"So it's all right for you to eat Indian food," I asked my daughter's friend after our second trip to the buffet. She looked puzzled. "You know that it is prepared with ghee or clarified butter, don't you?" The Hindus, I went on to tell her, refer to foods that are cooked with ghee as *pukkhā* or authentic, superior foods, and they refer to foods cooked with water or oil as *kachchā* or inferior cooked foods. They believe that the purity of the cow, imparted through its milk to cooked food, makes *pukkhā* foods superior. "But perhaps," I said, "it might also have something to do with the quality of the animal protein."

Now she looked horrified. "Oh no," she said emphatically, "not here. Not in this restaurant. Here they cook only with vege-table oils."

You gotta think what you believe, I once heard a basketball player tell a spectator who was taunting him from the stands. It was twenty-five years ago at a game in New Orleans, but I haven't forgotten his words. He was absolutely right. People do think what they believe—what they already believe. And as I'm learning, they also eat what they believe. I wanted to talk to my daughter's friend about the monkeys in tropical forests eating their wormy fruit. I wanted to tell her about vegans in India who get their vitamin B12 from the insects and bacteria that normally contaminate their foods, a curious piece of dietary trivia that was discovered only when a group of these vegans moved to England, where foods are often sprayed with pesticides and washed before consumption, and soon came down with pernicious anemia. But I let the subject drop. For who was I to talk about flexibility and the virtue of high-quality foods? I was the one who had tried to eat a salad every day in the Arctic.

Nine

THE ONLY COOKS ON THE PLANET

All the howler monkeys in the troop had left the carne asada tree where they had been feeding—all except one mother and her ten-day-old infant. She was sitting there, high up in the tree, her tail wrapped around a branch, her infant clinging to her belly, when suddenly she began turning in circles. She spun around twice, then lost her balance and toppled over. At first her tail, wound tightly around its branch, prevented her from falling, and she hung there for a moment upside down. Then she began having convulsions, and she and the baby plummeted thirty-five feet to the ground. Dazed but uninjured, she climbed back up into the tree and remained there quietly in one spot for the rest of the day. She didn't eat for the next twenty-four hours.

Others beside Katharine Milton have spent time observing howler monkeys in the forests of Central America, and one of them, Ken Glander, a graduate student at the University of Chicago, happened to be watching when that mother fell from the tree. Glander was in the northwestern corner of Costa Rica in a tropical dry forest when he saw this strange occurrence. It was one of the few times he has seen a howler fall from a tree, though he has since spent many years in this forest and has witnessed hundreds of death-defying, far-flung leaps between trees. "A tree is everything to a howler," explains Glander, who is now the director of the Duke Primate Center in Durham, North Carolina. "It is food; it is shelter; it is security. Howlers leave trees very reluctantly."

And yet Glander hadn't been entirely surprised by that fall many years ago, for there had been, in the weeks preceding it, a number of strange occurrences in the forest. Just the day before, a young howler from that same mother's group had disappeared while the group was sleeping in trees along the bank of a river, and Glander suspected that it too had fallen and had been so unlucky as to hit water and drown. In the two weeks prior to the mother's fall, six howlers from other troops in the area had been found dead. Two young females had been observed twitching and jerking on the ground.

In fact, the fall made some sense to Glander—as did the disappearance, the deaths, and the illnesses. He and his wife Molly had been watching these groups of howlers for twelve and a half hours every day, so they knew that at least half of the dead animals and all of the sick ones had recently been feeding on the leaves of either a carne asada tree or a madera negra tree, two trees that are common in Costa Rica's tropical dry forests. And he knew that the leaves of both of these trees contain potent chemicals that local Indians use to kill fish and rats. Indians crush carne asada leaves and put them in lakes or rivers to poison fish, and they crush madera negra leaves and mix them with water and rice or seeds to use as a rat poison. Carne asada leaves have been found to contain the alkaloid andirine, which, in very small quantities, will make humans extremely nauseous.

Madera negra leaves contain rotenone, the active ingredient in many insecticides. It looked to the Glanders as if the monkeys had poisoned themselves, and though they couldn't prove it, they could and did rule out many other causes of death. Extensive autopsies of the dead monkeys produced no apparent pathologies. Tissue and blood tests eliminated herpes virus, rabies, and yellow fever as the possible cause.

It was the German botanist Ernst Stahl who first suggested, as long ago as 1888, that not everything green is palatable; Stahl who first suggested that plants use invisible chemical defenses, in addition to their highly visible thorns and burrs, to protect themselves from being eaten. Despite Stahl's early insight into plant defensive strategies, it wasn't until the late 1950s that botanists began to take plant chemicals seriously as defensive weapons. And not until the 1970s that the true nature of these chemicals was finally understood. Thousands and thousands of the chemical substances that plants are known to produce—rotenone in madera negra leaves, nicotine in tobacco—are of no physiological importance to the plants whatsoever. Their only purpose is defensive. Their only job is to prevent a plant from being eaten. Thus they came to be called secondary compounds—compounds, in other words, with no primary metabolic role.

Secondary compounds probably began as waste products of a plant's metabolism, conferring, by some chance, protection against plant-eating microbes or insects. But they were transformed long ago into a plant's armed services. Plants cannot flee from hungry predators, of course, so they became experts in chemical warfare instead, creating for themselves arsenals of bitter-tasting and poisonous compounds. No one knows how many different secondary compounds there are, how many different alkaloids, tannins, phyto-hemagglutinins, oxalates, phytates, and cyanogenic and cardiac glycosides plants have dreamed up to foil plant-eaters. But more than ten thousand have been identified so far, and some researchers say that those are only the tip of the iceberg. Individual species of plants often contain numerous different kinds of these secondary compounds. Some of the trees in Glander's study area in Costa Rica have more than ninety.

But producing these arsenals is expensive for plants. It requires energy that any sensible plant would rather put into growth or seed production. So many plants have cunning ways of lowering their costs. *Plants produce these weapons only if they need them.* In Africa, for example, yams that are toxic to eat are found only out on the savannas or at a forest's edge. Yams inside tropical forests are perfectly edible because, it is thought, they are scattered among so many other plants that their risk of being eaten is very low. *Plants only arm those parts where predators might be feeding*—lower or higher branches, depending on the herbivore. Some plants can even change their chemical defenses in response to attack, with a response time ranging from forty minutes to several years. *Plants put their chemicals into the parts that count the most: mature leaves (the power plants, if you will, the energy producers of plants), seeds, and other reproductive parts.* For years I had been puzzled by the change that takes place in the watercress in my woodland spring as the plants grow tall and begin to flower, puzzled by the fact that in just a day or two, leaves go from tasting sweet and peppery to peppery and bitter. I used to think that watercress plants just "naturally" become bitter as daytime temperatures rise, but that bitterness, I now suspect, is protection for the plant's newly emerging flowers and seeds.

Despite the very large number of secondary compounds that are known to exist, these compounds act in two ways. Some disrupt the metabolism of cells and are toxic to cells, while others interfere with digestion. The toxic compounds, including the bitter-tasting alkaloids in tea, coffee, and tobacco, leave the stomach and the intestine through the bloodstream and enter cells, any cells—liver cells, lung cells, blood vessel cells—where they cause a whole host of effects. Nicotine, for example, damages the cells that line the blood vessels and the airways in the lungs, causing holes that may lead to artery disease or bronchitis and cancer. Other toxic chemicals enter cells and release cyanides and carcinogens, or they inhibit the enzymes responsible for the transport of ions across the cell membranes of heart muscle, or they bind essential minerals and vitamins, or they burst red blood cells. Pretty serious stuff.

All of our most familiar and most addictive drugs are toxic alkaloids (nicotine, as well as cocaine from coca leaves, morphine from the opium poppy, and cannabidiol from hemp). And while causing animals to become addicted to the substances in your leaves is one hell of a way to keep from being eaten, you might be thinking, most of these compounds evolved to protect plants against their earliest and most important predators: insects and microbes. Mammals are latecomers on a plant's defensive horizons, and mammals react differently to the same compounds that deter insects. The caffeine in the cup of coffee that gets you going in the morning is first and foremost an effective insecticide.

Secondary plant compounds that work by reducing the digestibility of foods stay in the stomach, and there they bind either to proteins (to make them unavailable to the digestive system) or, like the antitrypsin factor in soybeans, to digestive enzymes (to keep them from doing their work). Tannins are the most familiar of the protein-binding agents. They get their name from the role that these compounds play in leather processing, for tanning is the technique of using tannins to bind proteins in an animal's skin. By interfering with digestion, both types of compounds affect the weight gain of animals that ingest them and ultimately their growth. Steers fed a high-tannin sorghum gained less weight than those fed a low-tannin sorghum. Humans fed high-tannin beans gained half the weight gain of humans fed low-tannin beans. The difference is not one of nutrients, remember, but of the availability of nutrients. Tannins and antienzymes limit the nutrients available for growth. Over time animals have learned this and many have learned to avoid plants and plant parts (bark especially) containing large amounts of these compounds.

"The world is not coloured green to the herbivore's eyes," says Dan Janzen, a biologist who has devoted much of his career to the study of plant defenses and a professor at the University of Chicago when Ken Glander was a graduate student, "but rather is painted morphine, L-dopa, calcium oxalate, cannabinol, caffeine, mustard oil."

In the face of secondary compounds, all plant-eating animals, whether insects or primates, have two basic strategies available to them. They can be specialists or generalists. They can feed on just one kind of plant and become expert at detoxifying the compounds that this one plant contains (like the koala, which eats only eucalyptus). Or they can eat many different kinds of plants and develop nonspecific ways of dealing with all the different compounds that those plants contain.

It would be impossible for generalists to maintain specific detoxification systems for all the chemicals that they might encounter during the course of a day. But they can have general detoxifying systems, and they can reduce the amount of any one chemical that they must deal with by selecting plant parts that contain little or no chemical protection; or by eating no more than small amounts of any one food; or by ingesting earth to bind some of a plant's tannins and toxins.

Chimpanzees eat clays, often from termite mounds, daily or seasonally, and it is thought that this practice has a detoxifying function. Humans have definitely been known to consume clay and earth for this purpose. Indians in California used to prepare their acorns with clay so that the clay would absorb much of the acorns' tannins and bitterness and make them safe for consumption. Indians in the Central Andes could not have used wild potatoes as a food source without simultaneously ingesting large amounts of earth. Geophagy, as this is called, eating dirt or clay, is also common during famines, and in Europe, clay was euphemistically referred to as "mountain meal." Eating earth when one is starving fills the stomach, of course, and gives one a sense, however misguided, of being satiated, but it also detoxifies the only other foods that might be available during a famine: wild roots and bark.

Another way that generalists deal with secondary compounds is by making sure that they are well nourished and that they take in adequate amounts of protein, since proteins, after all, are the stuff of detoxifying enzymes. Among humans, malnourished individuals have decreased levels of detoxifying enzymes and altered abilities to metabolize drugs. Some studies of primates have found that monkeys and apes tend to avoid plant parts that contain high levels of secondary compounds. Others have found that these animals seem to be making a sophisticated analysis of the cost-benefit ratios of eating specific plants, accepting, it seems, a certain amount of poison for a high protein content. Researchers have concluded that there is a dynamic and complex relationship between the presence of secondary compounds in a plant and primate feeding behavior, a relationship that hinges on the type of defensive compounds present, the amount of nutrient present, and the digestive system of the primate. Given that individual plants contain varying amounts of the chemicals that their species are known to produce and that the only real way to tell exactly what chemicals a monkey or ape is ingesting is to watch it carefully, then analyze the specific plant parts it has been seen eating, it is no wonder that this relationship has been difficult to unravel.

For instance, of the 149 madera negra trees in the forest in Costa Rica where Ken Glander was conducting his field study of the feeding habits of howler monkeys, howlers usually fed on only three. And only these three, Glander found, had leaves that were free of alkaloids and cardiac glycosides, compounds with which the other trees were loaded.

"Imagine living in a town with one hundred and forty-nine restaurants," says Glander. "Most put poison in the food they serve, but three serve safe, healthful meals. The people of the town would learn quickly, by sad experience, which restaurants were safe."

Then imagine what happens if those three restaurants close or run out of food, and everyone is forced to eat elsewhere.

This was the howlers' situation in the year that Glander saw them spinning on the ground and falling from trees. Three years of drought had made the trees all over the northern part of Costa Rica cut back on their leaf production, so howlers, which are usually very capable of avoiding those carne asada and madera negra trees whose leaves are laced with toxins, had to eat poison if they were to eat at all. They couldn't leave the area and go elsewhere because howlers are confined to areas where there is a continuous tree canopy (and these areas are rapidly diminishing in size). Howlers don't like to leave the safety of trees, as we know, not even to cross a road or farm.

A nimals other than howler monkeys have poisoned themselves by eating things that they shouldn't, as Americans discovered when they first began to move to lands west of the Appalachians, and to settle the area of Pennsylvania, Ohio, Indiana, and Illinois that was once known as Transylvania. Abraham Lincoln's mother, who died when the future president was just seven, may have been the most famous person to succumb to the "milk sickness," an often-fatal disease that was eventually linked to drinking milk or eating meat from cows that had been foraging on a native plant, white snakeroot (the same plant that deer in my part of the country avoid). But in Danville, Indiana, one tenth of the population died of milk sickness in a single year. In Madison County, Ohio, one quarter of the population died; and in Dubois County, Indiana, as many as one out of every two people. So serious was the disease in some places that in 1836 a contributor to the *Transylvanian Journal of Medicine* warned that "some of the fairest portions of the West, in consequence of this loathsome disease, must ever remain an uninhabited waste, unless the cause and remedy can be discovered."

"The wilderness is careless," Carl Sandburg wrote of this epidemic in *Abraham Lincoln: the Prairie Years.* A person with the disease usually became extremely weak, nauseated, thirsty, and constipated, sometimes with a burning sensation in the stomach and always with a foul breath that smelled like turpentine or acetone. Doctors attributed the disease to different causes: water spiders, tsetse flies, poisons in the soil or water, but settlers quickly recognized that it occured in areas where cattle were suffering from trembles, a disorder named for the trembling that followed even their slightest exertion and whose symptoms also included "pungent" breath. In the new territories of Ohio, Indiana, and Illinois in the first part of the nineteenth century, cattle were dying of trembles by the thousands, and where they were dying, humans were coming down with milk sickness.

The cattle that developed trembles, it was also noticed, had frequently been feeding in the woodlands of the new, largely uncleared and unfenced territories. Early measures on the part of settlers to prevent trembles and milk sickness included fencing forested areas "to prevent animals from eating an unknown vegetable, thereby imparting to their milk and flesh qualities highly deleterious." The search was on for this unknown vegetable, and the hero of this search was Anna Pierce, an Illinois woman who had lost several family members to milk sickness. Pierce, who had taken midwife and nursing courses before moving west, made several important critical observations about the twin epidemics of milk sickness and trembles. She noted, for example, that the diseases were seasonal, occurring only during the growing season, and that horses, which eat more grass than leafy plants, are rarely affected. She also had the good sense to talk to a local Shawnee woman, who took her right to the broad-leafed herbaceous plant that she said caused both trembles and milk sickness, the plant that the Shawnee use to treat snakebite, the plant that botanists call either *Eupatorium rugosum* or *Ageratina altissma*.

When Pierce fed white snakeroot to a calf and it developed trembles, she had all the proof she needed to start a program aimed at eradicating this plant from the new territories. But it wasn't until 1928, more than one hundred years after the first settlers began dying from milk sickness, that the active compound of snakeroot, tremetol, was isolated and injected into test animals to successfully produce trembles. In the interim, many physicians would continue to look elsewhere for the cause of milk sickness, and one was so skeptical of Pierce's claim that he ate snakeroot—and promptly died of his disbelief. Tremetol is really a number of different chemicals, but the most abundant among them is very similar to rotenone, the same compound that the howlers had been ingesting when they resorted to eating madera negra leaves.

And like howlers, the cattle that had eaten white snakeroot would have preferred to be eating something else, but snakeroot remains green until late into the fall, and during dry years, it is often the only green thing around. If given a choice, though, cattle always eat something other than this noxious food. During wet years and in richer pastures, they remain healthy even if they are let out into areas where snakeroot is growing. Another mystery about snakeroot was solved when plants from different parts of the country were tested for the presence of tremetol. White snakeroot grows in many parts of the East and the South, and settlers in Transylvania wondered why they had had no trouble with milk sickness in their old homes. But the amount of tremetol in white snakeroot varies markedly from location to location because of soil quality or other conditions. The white snakeroot in my woods (plants that stayed green this year until early November) may not be as potent as it is in the Midwest, but it is still something to be avoided.

Perhaps I have given you the impression, from these stories of howlers and milk sickness, that plants often win in their battle against herbivores. That with their well-stocked arsenals of secondary compounds, they have driven away their plant-eating enemies. If so, then I have greatly misled you. The arms race, as it is often called, between plants and plant-eaters is ongoing, and animals are continually coming up with ways of outwitting a plant's defenses. One example is those caterpillars that protect themselves from light-activated toxins in plants by building a shady home of rolled leaves in which they can safely dine. Another is the parsnip webworm, which can eat the seeds of parsnip plants (seeds that are deadly to humans) because it excretes most of the toxins into the silk it uses to make its web. A third, and one that I will come back to, is the many cooking techniques of Homo sapiens.

And if I've given you the impression that secondary plant compounds are always to be avoided, I've also misled you. Secondary compounds were first produced by plants to protect themselves against attack by insects, fungi, bacteria, and other microbes. Other animals, though, may find these same compounds more useful than harmful. And this is not too surprising, since infections of fungi, bacteria, and parasites are problems for *both* animals and plants.

The widespread use of leaves as medicines and condiments in human diets suggests that, far from being avoided, leaves and other plant parts have been sought out for their chemicals—even though those chemicals may have toxic as well as beneficial effects. Much medicine is, after all, nothing more than controlled toxicity: killing more invading cells than your own; tolerating some side effects in the hope of a cure. William Withering pioneered the use of digitalis as a cardiac stimulant in Western medicine in the eighteenth century, and the dose he found to be effective was very close to the toxic dose, the dose that could kill a patient. Nicotine causes holes in blood vessels and other cells in the body, but

researchers have been finding that it is also helpful in the treatment of attention deficit disorder and Alzheimer's disease.

Using plants to cure or prevent ills is probably as old as plant-
eating itself. Diet and medicine, as physicians and healers in some
cultures seem to be very aware, are two sides of the same coin, and
Western medicine can, in fact, be seen as an attempt to replace
many of the secondary compounds that were once part of the normal diet. It used to be thought that the reason people in hot
climates eat lots of highly spiced foods is that food spoils faster
in those climates and spices disguise the taste and smell of food
that is going bad. But two Cornell researchers recently tested
numerous common spices, among them allspice, oregano, thyme,
cinnamon, tarragon, cumin, cloves, bay leaf, cayenne, rosemary, mar-
joram, and mustard, and found that all of these spices have
powerful antibacterial and antifungal effects. All are capable of
killing or suppressing seventy-five to one hundred percent of the
bacteria and fungi that commonly contaminate and spoil foods.
Onion and garlic are just as effective. Spices are most often used in
tropical and subtropical regions, these researchers argue, not
because they hide the fact that food is spoiling but because they
kill the organisms that are spoiling the food.

Eighty percent of the world's inhabitants still rely on their
knowledge of plants and their effects for their primary health care
needs, to treat or to prevent not only parasites and other problems
stemming from food contamination but also skin problems, pain,
wounds, diarrhea, headaches, fever, colds, eye problems, and even
cavities, since there are reports of various tannins inhibiting the
activity of oral bacteria.

Humans are not the only animals to appreciate the medicinal
effects of specific plants. And why not? If animals other than our-

selves have learned to avoid plant species or plant parts because they cause them harm, as they clearly have, why can't they learn that the ingestion of certain plants makes them feel better? Why can't they also learn to exploit the medicine chest of the natural world?

Dan Janzen, now at the University of Pennsylvania, was probably the first to suggest that *nonhuman* primates use plants medicinally when, in 1978, he linked the absence of intestinal parasites in black and white colobus monkeys in Uganda to their regular ingestion of plant secondary compounds. But recent reports of the highly unusual way that wild chimpanzees eat leaves from the genus *Aspilia* provide the most convincing evidence for self-medication in a nonhuman animal.

Normally when chimpanzees eat leaves, they stuff them into their mouths as fast as they can. But when they eat *Aspilia* leaves, which people in Africa use to treat stomach disorders, they take only one or two young leaves at a time, roll them between the tongue and cheek, then swallow them whole. Swallowing the leaves whole prevents whatever compounds are in them from being broken down in the stomach, delivering them intact to the small intestine. Field researchers who observed this curious behavior were intrigued and tested leaves from the same plants that the chimpanzees had been eating. They found that they contained high concentrations of antibacterial and antifungal compounds, as well as the potent antinematodal, or worming, agent thiarubrine. Worms may indeed have been the chimps' problem, since chimps were seen to ingest leaves in this way only during the season when the incidence of intestinal nematodes was the highest.

Another persuasive example of self-medication in chimpanzees is the case of a lethargic chimpanzee that had dark urine and diarrhea but recovered after sucking the juice from the stems of *Veronia amygdalina*, a plant known as "bitter leaf" to local people, who use it to treat worms, intestinal upset, and appetite loss.

Secondary plant compounds not only cure infections and kill worms. They can also change an individual's mental state. And plants have been sought for this purpose, too, for a long time, certainly as long as humans have been keeping any kind of records of their activities. Ancient use of psychotropic or mind-bending drugs has been inferred from mushroom-shaped stone carvings in areas where hallucinogenic mushrooms are still in use; from pipes, presumably for sniffing or smoking; and from portrayals of drug use painted on pots or carved in stone. The earliest concrete evidence of the use of psychotropic substances involves coca leaves associated with mummies in coastal Peru in 500 B.C. Alcohol, the *intoxicant* most frequently used by humans (the word *intoxicant* contains within it both the pleasures and hazards of these substances), is not strictly a secondary plant compound, but it is a compound derived from the fermentation of plants that is both mind-altering and toxic in excess. Its overconsumption can cause the same cirrhosis, or failure of the liver, as eating a poisonous mushroom, and its use dates back even farther than that of other psychotropic compounds, back to Hammurabi's elaborate laws about wineshops recorded on stone in Babylon in around 2000 B.C., back to ancient Sumerian tablets from 3200 B.C. listing the ingredients for beer.

So why do we seek out and ingest these dangerous substances? Why do we risk destroying our livers and our brains? To some extent, the answer might have to do with medicine and nutrition, since certain psychoactive plants have important therapeutic consequences (in the treatment of parasites, for example) and since fermented grains, as we know, are important sources of both calories and vitamin B12. But this can hardly explain the widespread popularity of these substances or the fact that alcohol and other substances have been used in every human culture known ever to exist. A few authors have suggested that some means of transcending reality is a basic need of all human beings, but what would those authors say to the fact that animals other than ourselves also seem inclined to ingest, and overingest, foods with mind-bending properties?

Migrating birds have been seen gorging themselves on fermented berries, then—drunk, disoriented, and overweight—the unlucky ones among them smash into buildings and cars. Pigs are very fond of marijuana, and ducks forage for a variety of narcotic plants. Reindeer on the Asian tundra ignore their normal diet of lichens whenever they smell the red-capped *Amanita muscaria* mushroom, a psychedelic used by Siberian shamans. Monkeys and dogs actively pursue opium smoke, and writer and opium smoker Jean Cocteau once observed that domesticated animals tend to form a "circle of ecstasy" around human users. Tobacco is highly addictive to a wide number of animals, including baboons, parrots, and Syrian hamsters; chimpanzees have even overcome their natural fear of fire in order to be able to smoke cigarettes on a regular basis.

"When given a choice of literally thousands of plant drugs in nature," says Ronald K. Siegel, a research professor in psychopharmacology at the University of California in Los Angeles and the author of a number of books on intoxication, "animals select the same types of substances regularly consumed by humans." And perhaps, he suggests, their reasons are not all that different from our own. Animals besides ourselves are vulnerable to stress: the stress of competitive feeding and competitive mating; the stress of group living; the stress of disturbances in the environment or, in artificial settings like preserves or zoos, of overpopulation or bore-dom. And like us, they may try to reduce that stress with alcohol and drugs. Much of the evidence for this interpretation of animal and human intoxication is anecdotal: free-ranging water buffalo in Cambodia forage on opium poppies and were observed to increase their foraging during the Cambodian War; consumption of fer-mented fruits by elephants in India seems to increase with population density. But one thing is certain: in the wild, animals have access to substances like alcohol and hallucinogenic plants only on a seasonal basis. They have little chance of becoming seri-ously addicted to these substances or of suffering the long-term physical and social consequences that human addicts suffer.

So humans are not alone in exploiting plants for reasons other than nutrition. But we are alone in treating plants to rid them of secondary compounds. We are the only animals to boil, dry, bake, microwave, soak, and otherwise process plants to enhance their edibility. We are the only cooks on the planet. We tend to think of cooking as a way of making foods taste better, but it has an even more important use. It is, first and foremost, a way of overcoming plant defenses, a way of making foods that would otherwise be toxic or indigestible com-pletely edible.

Meat is perfectly edible and digestible in its raw, unadulter-ated state, but in order for legumes, like beans and peas, to become the important sources of amino acids that they are today, humans had first to figure out how to remove their secondary compounds, including their potent antienzymes. The same is true of all of the major domesticated cereal crops. Heating and other cooking tech-

niques break down enzymes and increase the solubility of alkaloids, which can then be washed away. In order for bitter manioc
or cassava root to become the staple food of West Africa, humans
had to discover how to carefully press and cook this fleshy root so
as to remove its potentially fatal levels of cyanide. In order for people in the Andes to live on the wild potatoes growing there, they
had to learn how to eat them with earth *and* how to leach, freeze,
dry, and boil these tubers to help remove their toxic glycoalkaloids.

It is ironic that there are people who think that the healthiest,
most natural way to eat is to eat only raw foods, when so many of
the foods in our kitchens and grocery stores are there only because
we have found ways to cook them and, in cooking them, transform
them into edible substances. Claude Lévi-Strauss was right, but for
reasons he little suspected: there is something very powerful and
fundamentally human about the transformation of raw food into
cooked food. The majority of secondary compounds that humans
are able to deal with involve cultural rather than biological adaptations, cooking rather than the liver's detoxifying enzymes. These
adaptations are stored in the cultural beliefs and practices of various groups and transferred from generation to generation. Because
all other animals depend on the slower processes of genetic adaptation to overcome the toxic qualities of various plants, humans
have been able to displace other species in many places.

Humans have also figured out how to select for or to create
changes in plants that rid them of undesirable secondary compounds. Plants, as we know, produce varying amounts of chemicals,
and by cultivating seeds from plants with few or no toxins, we have
also developed increasingly edible strains of formerly less edible
foods. Yams have been rid of alkaloids and phenanthrenes in this
way; sorghum of tannins; lima beans of cyanogenic glycosides; potatoes of glycoalkaloids; eggplants of glycoalkaloids and saponins; and

pumpkins, cucumbers, and watermelons of the cucurbitacins that used to make them taste bitter. The obvious benefits of these new domesticates include new sources of calories and new environ- ments in which to live. With the introduction into China of the domesticated potato, a food that allowed the Chinese to settle in the mountains, the population doubled in size within just 150 years. The not-so-obvious disadvantage is that domesticated plants do not always supply the same amount of nutrients and beneficial sec- ondary compounds as their wild forebears.

Another problem with plant selection versus cooking is also not appreciated. Selection for less toxic, more digestible plants makes these plants edible to many animals. Cooking of toxic plants makes them available only to humans. Humans have little competi- tion for the roots of the cassava plant because of their high levels of cyanide. These highly productive roots are naturally immune to migratory African locusts and highly resistant to populations of wild pigs, baboons, and porcupines. But for our new and tastier strains of potatoes and eggplants, we have to compete with bacte- ria, fungi, insects, birds, and other mammals. Cultivation of these nontoxic varieties usually requires fences and pesticides—a new arms race between plants and animals, with humans, this time, weighing in on the side of the plants.

Ten

THE HUMAN OMNIVORE

I tell you this story for a reason.

I was a visitor in Charleston, South Carolina, one hot day in July, not too many years ago, when I happened to wander into the small park next door to the Charleston Historical Society. I don't remember what I noticed first—the large live oaks shivering with activity or the incredible stench—but both were explained by a sign that stood a few feet within the gate.

Welcome to Washington Square Park, the sign read. *Each spring the Yellow Crowned Night Heron arrive in Washington Square Park to build nests and rear their young. This process begins in April and extends through July. This activity provides a unique opportunity to view local wildlife in an urban setting. We apologize for any inconvenience the yellow crowned night heron may cause you. Please enjoy the park.*

On a bench inside the park, a man with crutches and a tobacco-stained yellow mustache was trying to do just that, though his first words to me were, "These birds stink!" His name was MacMullen, and he told me that he had been coming to the park for two or three years, and that the herons were not the strangest things he had seen there. Not long before, he had seen "a big yellow hawk go after a pigeon. When it was finished, there was nothing left—no bones, no feathers, no beak. Nothing."

Maintenance men from Charleston's department of parks arrived and began what I learned was a daily routine of hosing down the benches, the walkways, and the monuments. It should have been the cleanest park in town, but of course it wasn't. I asked one man with a hose how many herons there were in the twelve trees, but he didn't know. "These birds are a nuisance," he growled. "And it's not just their smell. The babies can't keep still, so they fall out of their nests. The ones that fall on the grass are the lucky ones."

I returned to the park at dusk that evening as the herons were preparing to leave for the night, and now the oaks were truly alive. Adult herons flew from tree to tree as young herons squawked in their nests and balanced on branches. One pair of adults left the park, then another. A third landed on top of a nearby building and stood for a while on pencil-thin legs, performing slow tai chi movements. It occurred to me that the herons must seem as strange to the other avians of the park—the giddy, cooing pigeons and the swift swallows (the city birds)—as they did to me. Everything about them was out of place: their slow, patient-angler movements, their large messy nests (overscaled for even these large oaks), their oversized babies. Watching them as the sky over Charleston turned electric blue, Maxfield Parrish blue, they seemed like one more animal that had lost its footing in a world order dominated by humans, one more thing that had gone wrong.

Imagine my surprise, then, when I got in touch with South Carolina wildlife biologist Priscilla Massenburg and asked her about the birds. "You can't appreciate how perfect the park is for the herons until you get up high, up above the trees," she told me. "I once went up in a cherry picker, and from that height you see water everywhere. There's the Ashley River off to one side of the park and the Cooper River off to the other, and both have extensive tidal marshes where herons find the fiddler crabs that make up most of their diet."

So, far from being a desperate choice of nesting grounds, Massenburg explained, it was as if the herons—thirtysix nesting pairs in all, plus seventyfive to eighty fledglings—had situated themselves between two giant supermarkets—both with their doors wide open and nobody at the cash registers.

And here's the reason why.

Sometimes the ways in which we humans go about finding food and eating also seem strange, even nonsensical, when viewed from one perspective or another: from a lunch line at McDonald's or Wendy's, for instance, or from within a psychiatric ward full of teenage girls who refuse to eat. But if we give ourselves some distance on these same phenomena, if we give ourselves *a bird'seye view* of humans and their relationship to food, a view that takes in not only humans but other animals, a view that considers the fact that males and females have different interests in food, a view that follows the human quest for food back in time, they begin to make a little more sense.

I'm not saying that the diets of humans always make *nutritional* sense, for clearly there is nothing healthy about anorexia or a highfat, fastfood diet. But rather, from a certain distance, from a perspective that includes what food means and has meant to all animals over time and that includes also how the human mode of foraging differs from that of all other animals, one can begin to see how people might make some of their food choices.

Humans tend to be very anthropocentric, for example, in their tendency to think that they alone have a complex and ambiguous relationship to food, that they alone have turned eating into a nuanced struggle between right and wrong. We assume other animals find eating simple. They eat when they are hungry and cease to eat when they are full. They eat the foods that are good for them, and they avoid the foods that are not. But what we don't appreciate is that *all* animals face essentially the same digestive task. All animals must distinguish between that which is edible in the world and that which is inedible. No animal eats everything in its sight. All animals must create a culinary universe, and for some this task is every bit as complicated as it is for humans.

For food specialists—animals that feed exclusively on one kind of food, like the panda bear that eats only bamboo, the parsnip worm that eats only parsnip seeds, or the carnivore that pursues only one type of prey—that task *is* relatively simple. Edible is everything that looks, tastes, and smells like bamboo, parsnip seeds, or that certain kind of prey. Inedible is everything that does not. Food specialists eat so few foods that the ability to recognize those foods—by taste, smell, and/or appearance—can be encoded in their genes. These animals run the risk of starvation if anything should happen to their well-defined food source, but they don't often eat the wrong food.

It can happen, though, as an incident in a Bronx apartment building makes clear. A Burmese python mistook his owner, a teenage boy, for a live chicken. Humans don't set off feeding responses in pythons, I was told by a herpetologist at the St. Louis

Zoo. But the python smelled the chicken that the boy had brought home to feed it, and motivated by the nearness of a favorite food, it attacked the next thing that it saw move. That's what pythons do—they locate their food on the basis of smell, not visual discrimination. In the wild, that next thing would have been the chicken, but in this apartment in the Bronx, it was the boy. The boy was found dead on the floor with the snake coiled around him. The chicken was found safe in its box. Though the lid was off and the chicken was free to leave, it had sense enough to stay out of the python's way.

So food specialists can make mistakes, but their gastronomic universes are pretty clear cut. For food *generalists*, though—the kind of eaters that most herbivores and all omnivores are—the task of creating a gastronomic universe is fraught with difficulty. Generalists are not dependent on any one food source and are better able to adapt to changes in their environments. But they have no way of coding in their genes the large number of foods that are safe to eat. There are just too many different kinds of edible substances out there. Plus, food generalists need to eat a diverse, varied diet in order to satisfy their nutritional requirements. (Omnivores differ from herbivores that are classified as generalists in that they need to eat both plants and animals in order to do this.)

But the more diverse one's diet is, the greater the chance that one might eat something harmful or poisonous—and the more cautious one must be. Animals that are indiscriminate about what goes into their mouths can easily poison themselves, as a handful of mushroom collectors find out every year. Then again, animals that are too cautious about eating can starve or become malnourished. Wild rats that are captured by humans are so suspicious of new foods that they refuse to eat anything they are given and quickly starve to death.

We think of omnivores as animals that can eat anything, as animals with no restrictions on what they eat. We think of the world as the omnivore's oyster. But eating is far too important an activity for this casual definition of omnivory. Jonathan Swift's "He was a bold man that first ate an oyster" comes far closer to the truth of what it means to be an omnivore and the disquiet with which all omnivores regard food.

In 1976 Paul Rozin, a professor of psychology at the University of Pennsylvania, first described this fundamental ambivalence about food, calling it the omnivore's dilemma or the omnivore's paradox, a conflict between the need to eat a variety of foods, on the one hand, and the fear of new foods, on the other. "Trying new foods is at the core of omnivorousness, but so is being wary of them," Rozin explains.

"Every omnivore, and man in particular," notes French anthropologist Claude Fischler, "is subject to this double bind, which probably results in various forms of anxiety." According to Fischler and Rozin and the many others who have accepted Rozin's fundamental insight into the nature of omnivory, much can be explained by this tension between an omnivore's neophobia (or resistance to change) and its neophilia (a need for change and a tendency to explore). This includes our particular vulnerability to concerns about food and food safety and the fact that people are more conservative about their diets when they are ill, or first thing in the morning when they may be feeling a little less adventurous, or as they grow old.

After childhood, most people show little or no tendency to experiment with novel foods, but during childhood, especially during the postweaning period, neophobia is greatly reduced. "In order to know whether a human being is young or old, offer it food of different kinds at short intervals," observed Oliver Wendell Holmes in

The Professor at the Breakfast Table. "If young, it will eat anything at any hour of the day or night. If old, it observes stated periods." Young children *must* be willing to try and accept many new foods because of their increased protein and calorie needs during a period of rapid growth and development. The fact that they are is consistent with the high incidence of poisonings in children this age. Interestingly, baboons have been found to exhibit a very similar pattern of behavior. In a study of wild baboons in Kenya that were forced to adapt to a new territory when farmers drove them from their traditional grounds, anthropologist Shirley Strum found that it was young baboons that led the troop in its search for new foods to eat.

Yet another manifestation of the omnivore's dilemma is the many complex rules that govern food consumption and eating behaviors in human societies and the fact that no culture is without its set of rules about food, its prescriptions on what and what not to eat. "Coping with or managing the anxieties generated by [the omnivore's dilemma] is not a process which goes on merely at the individual level," observes anthroplogist Solomon Katz. "Rather, solutions and coping strategies must emerge and be sustained at the cultural level." Cultures provide a way of "marking" what foods are safe to eat, and through cuisine, they make the necessary variety of foods available. "Given a choice," an old saying goes, "man tends to eat what his ancestors ate before him."

Cultures also endow certain foods with great symbolic meaning, and foods figure strongly in religious rituals, emblematic of important occasions, emotional states, or—in the case of the Christian tradition of communion, in which a wafer represents the body of Christ—important persons. Food, according to Katharine Milton, is an ideal symbolic medium, not only because it is essential to life but because there is some recognition, either conscious or unconscious, that it is both dangerous and powerful.

Because humans eat squirrel brains in Kentucky, ptarmigan dung in the Arctic, and duck embryos in the Philippines, we tend, as I've said, to think of ourselves as a species that eats just about anything. But of the very large number of foods that are available in any one location, we consume relatively few, more in areas of higher population density. Of the approximately fifty thousand edible plant species in the world, for example, the average American eats only thirty. This is not just the result of industrialization, as one might assume; it is true even in hunter-gatherer societies. As recently as the 1970s, the !Kung obtained ninety percent of their food from just twenty-three of the hundreds of species of plants and animals in their territory. And it is true of other omnivores as well. Chimpanzees at Kasoje, a western Tanzanian field-site in the Mahale Mountains, ignore the nuts of oil palms, but at Gombe National Park in Tanzania, they eat them with relish. Geza Teleki, a primatologist who spent many years observing chimpanzees at Gombe (the same chimpanzees that Jane Goodall made famous), wrote of his puzzlement over some of the omissions in the chimpanzees' diet there. For instance, they have never been seen to exploit any of the numerous fish, amphibian, and reptile species available in the park, and they even pass over fish being dried on the beaches—though those are eagerly eaten by baboons.

But all omnivores eat only a portion of the edible resources available to them. All omnivores are more conservative than not.

Besides the obvious appeal of foods that are high in fat and salt—two essential nutrients that were scarce in the African savannas where humans evolved, and for which humans now have a highly manipulatable taste—it has been suggested that part of the appeal of a McDonald's or a Wendy's is the familiarity of their foods and the predictability of their menus, so reassuring to an omnivore. And anorexia may be an extreme way of dealing with an omnivore's food anxieties, anxieties that are exacerbated in young girls because of hormonal changes and because of adolescent sensitivity to issues of body image and independence.

These are the kinds of things that one can begin to see with a little distance, from a bird's-eye view.

So how are omnivores equipped to handle this essential ambivalence that is at the core of their dietary behavior? They must have mechanisms or special behaviors that allow them to resolve the tension between neophobia and neophilia, mechanisms and behaviors that may sometimes go awry but that usually enable them to select a variety of appropriate foods.

Not surprisingly, they do. Omnivores are guided in their search for food by their sense of taste and smell (human taste receptors can detect sourness, sweetness, bitterness, and saltiness, four qualities of food highly relevant to our nutritional and biological heritage and to life in both trees and on the savannas); by hunger, that innate physiological need to eat that is experienced as a drive to obtain food; and by the two cardinal rules of survival for any hungry omnivore: (1) Eat nothing that tastes unpleasant or smells disgusting; and (2) Eat only a small amount of any new food, then wait to see if it's agreeable to your stomach.

Omnivores are exquisitely sensitive to unpleasant smells and to the consequences of the different foods they eat. They form long-lasting aversions to foods that they associate with stomach illness, even though those foods might have been eaten hours before the illness occurred and might not be responsible for the illness, an unfortunate fact of life for those undergoing chemotherapy and finding themselves associating their nausea with whatever food they happened to eat before treatment. John Garcia, the UCLA psychologist who first discovered how biased omnivores are toward remembering the foods that make them sick, likes to relate the story of how his mother came to detest chocolate. At the age of three, just before going on a boat trip on which his mother became seasick and vomited, she ate several chocolates. Intellectu- ally, Mrs. Garcia knows that chocolate doesn't cause seasickness, but try telling that to her omnivore's stomach and brain. She hasn't eaten chocolate since that trip.

Mrs. Garcia's aversion was to chocolate, but the greatest per- centage of food aversions (at least among North Americans who have been polled on the subject) are to protein-rich foods, particu- larly meat. Some researchers have suggested that this is a reflection of the importance of hunting and scavenging in early human evo- lution, but rats, too, develop aversions to proteins but not to carbohydrates—and rats aren't hunters. So perhaps a better expla- nation for what seems to be an innate predisposition to form aversions to meats is that protein-rich foods provide a rich medium for the growth of bacteria and other dangerous organisms. Eating plants can also be dangerous, but the problems posed by plants are not usually as insurmountable as those of meats. One bite of con- taminated meat may cause vomiting and diarrhea, but one bite of a toxic plant (except in the case of certain mushrooms) can be dealt with quietly and internally by the liver's detoxifying enzymes.

The Garcia effect, as it is now known, allows omnivores to learn about the negative effects of new foods, particularly proteins. But what keeps omnivores trying new foods? Neophilia, after all, is just as important to the survival of an omnivore as neophobia, and the omnivore that eats a wide variety of foods is much less likely to develop specific nutritional deficiencies. In nationwide studies, people who eat the most fruits and vegetables are half as likely to get cancers of every type as those who consume the least. In Japan, "it is common knowledge," two elderly Japanese once told me, "that you need to eat thirty-one different kinds of foods a day." Most humans get their vitamin B12 from animal foods and their vitamin C and many essential fatty acids from fruits and leafy vegetables. But what motivates human omnivores to eat all of these different things?

"Pleasure," says Barbara Rolls, director of the Laboratory for the Study of Human Ingestive Behavior at Pennsylvania State University. "The quest for pleasure, the fact that the perceived pleasure—the palatability—of a food declines as it is eaten while the perceived pleasure of foods that have not yet been eaten stays high."

"In compelling man to eat that he may live, Nature gives an appetite to invite him, and pleasure to reward him," observed the nineteenth-century gastronome Anthelme Brillat-Savarin, the same insightful Frenchman who coined the aphorism "Tell me what you eat, and I will tell you what you are," and my favorite, "A meal without cheese is like a beautiful woman with one eye."

Perhaps you have noticed the phenomenon that Rolls has been studying for the past twenty years, the phenomenon called sensory-specific satiety. You sit down to dinner, and everything smells delicious. You can't wait to try the swordfish, the rice pilaf, and the sautéed green beans. But after several minutes and several bites, that fish, rice, and beans don't seem quite as desirable, or palatable, to you as they did before you started the meal. This is not just a question of feeling full. The salad and the cheese that you haven't tasted yet remain just as desirable, as does the chocolate soufflé that you're anticipating will end the meal. Plus, in numerous experiments with normal-weight individuals, Rolls has discovered that this decrease in perceived palatability occurs whether or not a food is filling. She has found the same decrease for foods that are low in fat or made with artificial sweeteners as for foods that supply regular amounts of calories. "There is a natural tendency for us to switch between foods," Rolls told me when I called her at her office to talk about her work, "and that tendency, driven by these changing palatabilities, helps us to eat a varied, healthy diet."

Omnivores other than humans also have the same tendency, and the very first mentions of sensory-specific satiety in the scientific literature refer to the behavior of chickens and rats. "When only one kind of food is offered [to chickens]," a psychologist noted in 1934, "satiety very soon occurs, but this can be retarded by mixing the food with one or two other kinds." In 1940 it was reported that the feeding preferences of rats could be reversed by controlled prefeeding. Generally, when given a choice between sugar and wheat, rats prefer sugar, but if they are fed sugar right before being tested, this preference reverses.

Sensory-specific satiety assures that omnivores eat a variety of foods, but it also has another important role in their feeding behavior. It encourages them to take in enough calories to meet their energy and nutrient needs. Human subjects, Rolls has found, eat as much as sixty percent more when they are presented with a wide variety of foods as when their selections are limited, a finding of some significance to those of us who are weight conscious and trying to cut back on food consumption. When Rolls asked her subjects to eat until they were full but offered them only one type of sandwich or pasta, they ate limited amounts. When she offered those subjects cream cheese sandwiches that differed in flavor or pastas that differed in size and shape, food intake increased, but by a modest fifteen percent. But when she offered them a full four-course meal consisting of a series of very different foods, their intake more than doubled.

"Meal patterns throughout the world stress the importance of the different sensory qualities of foods," Rolls explains. "It just so happens that the way meals are set up in many cultures is to maintain as high a degree of palatability as possible. A meal of soup followed by meat with vegetables and ending with dessert, with an emphasis on contrasts in color and texture, reduces the possibility of the appetite diminishing during the meal."

"Personal experience," adds this physiologist, who described herself to me as a restrained female, someone who is very aware of what she eats, "indicates that it may require a very large amount of food leading to feelings of nausea and bloatedness before one refuses to eat at least small amounts of a particularly delicious sweet."*

---

*When Paul Rozin first introduced the concept of the omnivore's dilemma or paradox, he and his former wife Elizabeth Rozin suggested that the various spice or flavor complexes of the world's cuisines—soy sauce, rice wine, and ginger root in China; olive oil, lemon, and oregano in Greece; wine vinegar and garlic in northern Italy—play a role in resolving this dilemma. Wrapping new foods in familiar spices facilitates the introduction of new foods into a culture or family, they proposed. But in light of recent research showing that spices have powerful antibacterial and antifungal effects—and in light of Rolls's work—spices might better be understood both as a way of protecting food from contamination and as a way of creating variety in a diet so as to encourage the consumption of adequate amounts of food. This view is supported by the fact that spices are used most in countries with hot climates, as we know, and in countries where individuals rely on bland carbohydrates—corn, rice, wheat, and other grains—to satisfy the bulk of their calorie needs. Our taste receptors may initially tell us to reject spicy foods like chili peppers, for the secondary compounds in many spices could be toxic to human tissues. But our ability to learn from experience allows us to use spices for medicinal and nutritional purposes *and* to add variety to our diet.

Rolls has found evidence of sensory-specific satiety in experiment after experiment with normal human subjects, and she has also found interesting variations on normal patterns in patients with bulimia and anorexia. Patients with bulimia nervosa express little decrease in the pleasantness of a food as it is consumed, which might help to explain how they can binge on a single food. Patients with anorexia nervosa, on the other hand, express a sharp decrease in the pleasantness of a food after consuming only very small amounts. "Obviously, this is not the only thing going on with these patients," cautions Rolls, "but it is a finding that may help us to understand the psychology of eating disorders."

Electrophysiological studies are also beginning to reveal how this phenomenon of sensory-specific satiety is built into an omnivore's brain. In studies of monkeys, researchers have measured the electrical activity of single nerve cells in the lateral hypothalamus, an area of the brain that motivates feeding in both monkeys and humans. When a monkey is hungry, nerve cells in the lateral hypothalamus fire at the sight or the taste of food. As the monkey consumes a particular food, however, the cells gradually become less responsive and cease firing altogether. Then, when the monkey is offered a new food, they begin firing again—mirroring exactly the phenomenon Rolls has been learning about from her human subjects. So variety is more than just the spice of an omnivore's life. It is the stuff of an omnivore's life, and the quest for it is built into the omnivore's brain. "New meat begets new appetite" is a saying once common in England, but what medieval Englishman could ever have guessed how true, how electrophysiologically correct, this saying really is?

The need for variety is so much a part of our physiological makeup that human omnivores create it wherever they are, sometimes out of very little. In the Arctic, where meals used to be meat,

meat, and more meat, Copper Eskimos had strict rules that prohibited cooking land and sea animals together, though both could be
served as different courses during the same meal. Frozen raw venison, for example, might be followed by boiled seal meat and then by
such delicacies as boiled marrow bones or raw seal intestines.
There, where so few foods were available, the degree of cookedness
was used to add variety to the diet. In a refugee camp in Ethiopia in the
1980s, where starving refugees were provided with small amounts of
millet, beans, and milk, they often traded those foods with local people
for foods that provided fewer nutrients and fewer calories but that
gave them a little something different to put in their cooking pots.
"People tend to lose track of the idea that pleasure is central to eating
and that variety is central to maintaining pleasure," Rolls says, "and it
doesn't matter whether you are eating in a refugee camp or a five-
star restaurant."

Eleven

CRAVINGS

Sensory-specific satiety is one of an omnivore's most important tools, but as Barbara Rolls would be the first to admit, it is just one of its tools, just one of the things that are going on during a meal. It is largely a short-term phenomenon (after a period of hours, Rolls's subjects report a renewed interest in the same foods that had fallen into their disfavor), and omnivores must adjust their food intake to their nutritional needs over the course of long lives.

To do this, animals need a certain amount of nutritional wisdom or good food sense. You might not know it from standing in that same lunch line at McDonald's or Wendy's, but humans, like other animals, come into the world with a certain innate knowledge of what and how much to eat. Take the fact that most of us, quite naturally, keep our protein intake low and eat enough carbohydrates (and fats) to spare that protein for body building and the other things that proteins do best. Or take our energy needs, the amount of calories that we need in order to supply our bodies with the energy to mow lawns, run marathons, write books, and teach mathematics. A person easily consumes one ton of food and fluid during the course of a year, and a daily excess intake of only one hundred calories (the amount of calories in a single pat of butter) has the potential to increase fatty tissue by twelve pounds after just one year, by 120 pounds after ten. Though it is apparent that the incidence of obesity in this country is rising, it should also be apparent that most people have a remarkable ability to adjust their energy intake to meet their energy needs.

"These smells are pleasant when we are hungry, but when we are sated and not requiring to eat, they are not pleasant," Aristotle noted many hundreds of years ago. The pleasure we derive from food is not a fixed property of food. It changes over the course of a meal, as Rolls has found, and it changes with bodily needs, whether those be for calories or for specific nutrients. One of the most fundamental jobs of the brain is sensing the internal environment of the body and assessing the levels of blood sugar, insulin, salt, and any of a number of other compounds. If the brain detects a deficiency, the body's sensitivity to its environment gets transformed into motivated behavior, into the desire for foods containing those substances.

"A given stimulus can induce a pleasant or an unpleasant sensation depending on the subject's internal state," the psychologist Michel Cabanac restated Aristotle in 1971. Cabanac dubbed this phenomenon alliesthesia and suggested that it could explain the specific appetites or cravings experienced by people with certain diseases or nutritional deficiencies, cravings for sugar in diabetics and for salt in people with hypertension.

Most poignant are the children who develop these cravings. In Clara Davis's famous studies on food selection in newly weaned infants, there was a nine-month-old boy who was diagnosed with rickets, a nutritional disease that results in softening of the bones and is caused by inadequate consumption of vitamin D. This very young boy elected to drink cod liver oil, but when he was over his rickets, he refused the fishy drink. An often-cited case concerns a one-year-old baby who craved salt and was found to have a tumor in his adrenal cortex that caused his body to excrete this necessary substance. Long before this baby boy learned to speak, he made his craving for salt clear to his parents by pointing at the salt cellar and refusing to eat unless salt was added to his foods.

"At eighteen months he was just starting to say a few words, and salt was among the first ones," his mother told his doctor. The boy's pediatrician, wisely, advised the parents to indulge the boy's craving, but when he was admitted into the hospital for testing, hospital nutritionists refused to add salt to his diet. As a result, in order to keep his salt concentration at the proper level, his body began to excrete water and the boy quickly died a painful and tragic death of dehydration. Other children with adrenal cortex deficiencies develop cravings for licorice and consume it in large amounts if given the option. No one ever told them that natural licorice contains the sweetener glycyrrhizin, a compound that helps to maintain mineral and water balance. But somehow these children come to associate the taste of this black candy with an improved sense of well-being.

Perhaps not every craving reflects a real physiological need. Not the cravings for pizza or hamburgers experienced by overfed, overnourished individuals. (Nutrition researchers in this country don't even like to use the word *craving* because it is so often used as an excuse for overeating.) But cravings are sometimes dismissed when they might, in fact, have an important purpose. Many of the cravings and taste aversions associated with pregnancy are dismissed as the by-products of a hormone-induced sensitivity to taste and smell. During the first trimester, though, many women find previously acceptable, bitter-tasting foods like coffee to be suddenly unpalatable. Bitterness is a good predictor of toxicity, and so it would make good biological sense that pregnant women would be better-than-normal poison detectors. I was able to give up smoking when pregnant with my first child only because I became nauseous every time I lit up.

Most specific hungers or cravings seem not to operate via special senses that allow us to detect a nutrient before ingesting it, but rather by delayed learning of the consequences of eating a certain substance, the postingestional consequences. However, the actual number and kind of taste receptors that humans possess is still a matter of some debate. There are four accepted kinds of human taste receptors, for sweet, sour, bitter, and salty substances, but different researchers have been questioning whether we might also have taste receptors for fats, carbohydrates, and certain proteins. Japanese scientists, for instance, have proposed adding to the accepted four a taste receptor for a flavor they call umami. Umami, whose name derives from the Japanese word for *deliciousness*, is associated with foods like meats, fish, mushrooms, cheese, and certain vegetables and with the glutamate amino acids. Several umami substitutes, including monosodium L-glutamate or MSG, have been used in Japan to improve the taste quality of food for longer than nutrition has been a science.*

---

*The question of taste receptors for starch is an interesting one. Rodents exhibit preferences for pure starch, but the finding that most humans do not display a preference for starch-derived polysaccharides and find the taste of pure starch very bland, would seem to conflict with the fact that starchy plants—rice, corn, potatoes, and cereals—are staples in the human diet. These staples, though, are rarely eaten raw. They are usually cooked

Humans, of course, are not alone in their manifestations of nutritional wisdom. Sheep don't come into this world with any preference for real sugar over artificial sweeteners like saccharin, but after being given flavors paired with either glucose or saccharin, they soon come to prefer the combinations that give them real energy. Over time lambs will decrease their intake of foods that are deficient in essential amino acids, and cattle will develop a preference for supplemental protein blocks when they are ingesting forage that is low in protein. White-tailed deer browsing among the shrubs in my garden like best those plants that I have carefully sprinkled with fertilizer, those plants that provide them more nitrogen than my unfertilized plants. They show their nutritional smarts every time they take a bite out of my vision of a lusher and more beautiful garden.

But wait a minute, you must be saying to yourself. Just how infallible can nutritional wisdom be, when almost all of the major diseases affecting people in countries like the United States today are related to diet? When a college-age friend of my daughter's ate nothing but hamburgers and buns and soon came down with scurvy? Just how smart can animals be about their diets when dogs will poison themselves on chocolate and sheep will stuff them-

---

and flavored with spices and sauces. Thus, say Merryl Beth Feigin, Anthony Sclafani, and Suzanne Sunday in an article on species differences in polysaccharide and sugar taste preferences in *Neuroscience and Biobehavioral Reviews*, rodents may have acquired an appetite for starch through a biological adaptation, while humans have acquired it through a cultural adaptation, that is, cooking and cuisine.

That said, perhaps some subpopulations of humans *do* have taste receptors for polysaccharides, since pure starch is readily eaten by some people. Amylophagia, as it is called, is usually considered to be a form of pica—a craving for unfit foods such as earth and chalk—but it may also relate to our need for a consistent supply of glucose.

selves on the ovine equivalent of a candy bar? In a study that is often cited as evidence against the notion of nutritional wisdom, pregnant ewes were given access to a variety of foods, including a concentrated source of carbohydrates. Every ewe in the study ate too much of this foodstuff and wound up with a higher carbohydrate-to-protein ratio than was good for them. The ewes became fat at first, and then listless, and most wound up miscarrying their lambs. How to explain these findings?

The same experiments of Clara Davis that advanced the idea of nutritional wisdom in children also help us to put nutritional wisdom in perspective. Davis, a pediatrician who studied the food choices of newly weaned infants in the 1920s and 1930s, found that young children are perfectly capable of selecting a very healthy, nutritionally balanced diet all on their own (they keep their protein consumption, for instance, at a healthy and consistent seventeen percent), when they are given a choice of wholesome, *unsweetened* foods. No one knows for certain what would have happened if sweetened foods had been included in these children's choices, since such an experiment has never been—and probably never will be—conducted. (What parents, ask yourself, would allow their children to take part?) But it is widely assumed from Davis's studies that if nutritionally poor foods were added to the choices, children would be unable to maintain adequate nutrition. Children, as any parent knows, show a particularly strong preference for sugar-containing foods. Plus, other animals with an innate taste for sugar have a hard time maintaining adequate nutrition when sweetened foods are added to their diet. Rats have an extraordinary ability to regulate their energy intake in accordance with their energy output over long periods of time, but only if the food that is available to them has no flavoring agents, especially sugars. You wouldn't think that a squirrel would find anything more delicious than a fat acorn, but given a choice between an acorn and a chocolate chip cookie, the squirrel will usually take the cookie.

And as stupid as these choices seem, one can't really blame them on a lack of nutritional wisdom. During the course of evolution, squirrels, sheep, and humans have rarely encountered large quantities of concentrated, high-energy foods. Young men were never tempted by an endless supply of hamburgers.* Why should the food selection mechanisms of animals include protection against overeating these things? Our human tastes for foods evolved and enabled us to survive in the forests and the African savannas where animals were lean and fibrous, food shortages were a fact of life, and sugar came only in the form of ripe fruits and honey, foods that were available on an intermittent, seasonal basis. The human sweet tooth, which once enabled us to find quick, occasional bursts of energy for our large brains, now causes us to fill up on energy-rich but nutrient-poor foods.

We think our predilection for sugar is so natural that it is shared by every living creature. But a taste for sugar is an advantage only when one's diet includes honey and ripe fruits. Cats and other strict carnivores aren't at all tempted by a candy bar or a bowl of sugar water. They don't have the enzymes that allow them to break down carbohydrates efficiently. Instead cats, even satiated

---

*The case of my daughter's friend, an American college student who came down with scurvy, is in fact a little more complicated. College cafeterias are normally places where orange juice runs like water, but this student didn't have a meal plan because he wanted to save money. He found that one of the cheapest foods that he could buy was large packages of frozen, precooked hamburgers on buns, of the kind made by White Castle. He wanted to be able to live on this diet, he *believed* that he could live on this diet, and didn't listen to his cravings for alternative foods until his gums began to bleed, an early sign of scurvy.

cats, respond avidly to foods suffused with the odors of meat or mixtures of amino acids. Not even all primates like sweet things. Those that feed on unripe fruits (beating, therefore, other primates to the punch) much prefer sour things. Then there are animals that have an even greater preference for sugars than we do. Hummingbirds and many insects are able to detect sugar at far lower concentrations than humans. We have sugar receptors on our tongues and in our mouths, but insects that imbibe sugar from plants also have sugar receptors all along their feet and legs.

The human taste for sugar may not be as great as that of some insects, but it can be a disadvantage even in parts of the world very similar to the original environments in which humans evolved. I learned this recently on a trip I took to Africa with my family and some friends. There I met a Hadza hunter who was part of a tribe that lives in the acacia savannas of northern Tanzania. He was in Arusha recovering from a puff adder bite that he had received while tracking for a private hunting safari, and when we were introduced, I was immediately struck by his direct manner, his easy good humor, *and* his black teeth. I wasn't surprised when he told me that meat and honey were his favorite foods, but I was surprised by the quantity of honey that he said he consumed each morning. Three mugs, he told me, holding up his fingers to show me that each mug was the size of a small glass. That was his breakfast, a meal that usually lasted him until evening. American breakfast cereals, which I had always thought to be loaded with sugar, suddenly seemed very restrained.

Since he enjoyed honey so much, I asked Abedi Shimba if he had ever considered keeping hives. "Why keep hives," he replied in Swahili (we spoke with the help of a translator), "when there is so much honey around?" I remembered the wooden pegs that I had seen in the trunks of baobab trees when we were driving through parks in northern Tanzania. They were put there, I had been told, by the Hadza so that they could climb for honey.* I was also told that the Hadza are often led to honey trees by the honey guide, an African bird that has the unique ability to digest beeswax but cannot break into hives by itself because of its small beak. So this is a bird that makes its living off the human sweet tooth, a bird that guides humans to beehives, then feasts on what remains after the hives have been raided.

The honey guide is probably man's oldest surviving partner in predation, much older than the dog or the falcon. It attracts attention by calling to a passerby with a loud chattering cry, like the sound of a shaken matchbox. Then it flies off a short distance and resumes its calling, eventually bringing a willing human to one or more hives. The only other animal that the honey guide leads in this way (though it tries to tempt chimpanzees and baboons) is the honey badger or ratel, a chunky, lumbering animal that not only can climb trees and break into hives but can also immobilize bees with its anal secretions. It is not known whether humans first observed and imitated the ratel's association with the honey guide or whether the bird initiated its partnership with humans. But the fact that some Africans stimulate the honey guide into action by mimicking a ratel's grunts and its habit of knocking on trees gives weight to the former possibility.

---

*The answer was so reminiscent of the answer Richard Lee got when he asked the !Kung why they didn't grow crops that I had to smile. "Why should we plant when there are so many mongongos in the world," Lee was told, an answer that became famous in anthropological circles. The mongongo nut is the linchpin of !Kung existence, providing them with most of their calories.

Abedi told me that the Hadza are superstitious about honey guides and that he believes that if he doesn't reward a bird for its efforts, the next time the bird calls, it may lead him to a black mamba, the most poisonous snake in Tanzania, or to the unpre-dictably dangerous cape buffalo. But the Hadza are greedy about their honey, and Abedi also said that he will let a bird lead him to three separate hives before he gives it its due, a piece of honey-comb. Do they also give the bird the bee larvae? I asked. No, the larvae are taken back to camp for the women to cook. Besides their easily digestible sugars, honeycombs contain useful amounts of pro-tein and fat in the form of larvae, pupae, and pollen.

All the Hadza have rotten teeth, I was informed, though Abedi tried to assure me that the state of their teeth had nothing to do with their consumption of honey. But what else were they eating that could cause that kind of decay? The Hadza *do* prepare their evening meal by boiling their various meats, vegetables, and tubers, then mashing everything together, but the desire to eat soft, mashed foods could be the result of existing tooth decay, as well as its cause. Abedi also said that his people prefer paler, lighter meats because they are easier to chew.

I wondered whether the Hadza have always eaten honey in such great quantity or whether this consumption had something to do with their current situation. Hunting is technically illegal in their territory, part of the Serengeti National Park, and must be done somewhat on the sly. Many game wardens turn a blind eye to the hunting of hunter-gatherers and make a distinction between those activities and poaching for profit or hunting for food by other local people, but the hunter always takes a risk. So perhaps Abedi and other Hadza men devote more time now to searching for honey because sugar just rots their teeth. Shooting an animal with one of their poison arrows could land them in prison, an experience, I was told, from which no Hadza man emerges with his spirit unbroken. Whatever the reason, though, the Hadza are no more able to resist the temptations of sugar than children in a candy store.

Concentrated salts and fats were also hard to find out on the savannas, and our preference for salty-tasting substances now causes some of us to ingest salt at ten times our nutrient needs. Fats, essential to the structure of every cell, are also easy to overingest even though we do not seem to have specific taste receptors for fats and even though fats, by themselves, are perceived as unpleasant. Combined with other foods, though, fats are highly desirable. We have a "fat tooth," says Adam Drewnowski, director of the human nutrition program at the University of Michigan, that is even stronger than our sweet tooth. The problem with fats is twofold. Since fats contain twice as many calories per volume of food as proteins or carbohydrates, one who eats high-fat foods until the point of feeling full can easily take in much more energy than needed. And high-fat foods are highly palatable. Because of their creamy mouth feel, they taste much better than low-fat foods. Again, we are not alone in feeling this way. Rats provided with a high-fat "supermarket diet" of chocolate chip cookies, marshmallows, condensed milk, milk chocolate, salami, peanut butter, cheese, and bananas quickly overeat to the point of obesity, gaining 269 percent more weight than rats fed only laboratory rations.

Yes, animals are nutritionally wise, but humans today can't count on that wisdom to navigate us through the shoals of modern-day life. Foods of the past were bulky and low in calories. Foods of the present are concentrated or dense and full of calories. A four-ounce chunk of roasted venison that our ancestors might have enjoyed around an open fire provided about 130 calories. The meat alone in a Quarter Pounder at McDonald's provides more than three hundred. And despite their high caloric value, many of the foods we eat today are low in nutrition. Many modern diets are inadequate in fiber, vitamin C, calcium, and other essential nutrients.

Variety by itself is usually not enough to produce significant weight gain, as researchers have found. But variety plus high fat and energy content is the downfall of many. To eat right and maintain a healthy weight in an environment where abundance, not scarcity, is the norm, an environment where food manufacturers can manipulate the fat, salt, and sugar in their products to the point where they have us by the evolutionary short hairs, we must also be aware of what we are eating, aware of the nutritional and energy content of foods.

In the United States, cravings for foods are largely individual affairs, but in other parts of the world, shortages of particular kinds of foods can cause communal cravings—widespread hungers for the missing nutrients. Meat hunger, a craving for animal protein and fat that occurs even when there are abundant vegetable foods to eat, was first reported by anthropologists working in sub-Saharan Africa. It has since been noticed wherever animal protein and fat (including milk, butter, yogurt, and cheese) are available only on a seasonal basis. In the desert of Western Australia, the search for *kuka*, meat and all fleshy foods, is a major preoccupation of the aborigines. In the forests of Nicaragua, good times for the Miskito Indians are when there is an abundance of meat; bad times, when meat is lacking. Miskito women frequently cajole and taunt their men to go hunting, and the men go, knowing that the women will not sleep with them if there is no meat in the village. Many societies where meat hunger has been observed, like the Yanomami of the Amazon basin, have two different words for hunger: one means "an empty stomach," and the other, "a stomach that craves meat."

It used to puzzle anthropologists as to why the Yanomami, who are forest gardeners or agriculturists, abandon their well-tended plots of plantains and go on prolonged treks through the forest in search of animals to kill. Underlying this puzzlement, reflects Kenneth Good, an anthropologist at Jersey City State College who spent twelve years living with a group of Yanomami, was probably "the ethnocentric notion that social progress consists of having a settled life, and that trekking is a kind of enforced hardship to acquire food in lean times." Times weren't lean, though, when the Yanomami packed up all their possessions and cleared out of their communal shelters. They had enough plantains and other vegetable foods to meet all their energy needs. But what they didn't have, Good discovered, was enough protein and fat. The Yanomami trek to obtain a balanced diet. On treks, hunting yields are double what they are near the Yanomami gardens where prey populations have been greatly depleted.

One needs a little reminding of meat hunger today when one so frequently hears the suggestion that humans should no longer raise meat for consumption, that it is far more efficient and less expensive to produce vegetable protein. Yes, one can obtain all the amino acids necessary by consuming a variety of plant foods. But meat is more than just protein; it is minerals, vitamins, and fats too. And though fat is a dread word in twentieth-century America, fats, saturated and unsaturated, are essential components of the membranes that surround every cell in the body and the chief constituent of the human brain; they are necessary for hormone production and carry with them the four fat-soluble vitamins A,

D, E, and K. Many essential fats or lipids are found in plant materials, but some are more readily obtained from meat, fish, or milk.

It is fraudulent, as English physiologist M. A. Crawford says,
to suggest that vegetable protein and vegetable oils can be substi-
tuted for all animal foods—fraudulent and developmentally
hazardous. "Protein is mainly concerned with body growth, but
lipids with brain growth," Crawford writes, " . . . and diets deficient in essential lipids produce an irreversible reduction in brain
size over a number of generations and irreversible impairment of
the learning ability."

Clearly, Americans don't need as much fat or meat as they tend
to consume, especially adults. Clearly, we need to stick to guidelines
recommending that an individual's fat intake be no more than thirty
percent of daily calories and that the majority of those calories be
from unsaturated fats. But fats may be an even more important constituent of meat than protein. A strictly vegetarian diet can be
healthy with careful planning or with fortified foods and vitamin
supplements (though there is some evidence that children raised
exclusively on such diets have slowed growth and development),
but meat has long been an important constituent of the human diet.
Is it surprising that while the world's population has doubled in the
past half century, its appetite for meat has quadrupled? Or that the
consumption of grain by livestock is rising twice as fast as the consumption of grain by people? Even in India, where the concept of
ahimsa, or noninjury to all living creatures, is widespread and
respected, the highest-income groups consume over seven times
more animal protein than the lowest.

I spoke recently with a person who had experienced meat
hunger. He was neither an African from the sub-Sahara nor a poor
farmer from India, but, rather, a performance artist from California
who had just finished a project called "living from the land: a 30

day performance." Mark Brest van Kempen, who is known for using the outdoors in his pieces, decided that he would pitch a tent in an unpopulated part of Colusa County, north of San Francisco, and live for a month entirely off the land. He brought with him nothing but his camping equipment, a fishing pole, a shaker of salt, and a video camera with which he intended to record his adventures and his search for food.

The experience was more than the thirty-eight-year-old somewhat-experienced outdoorsman had bargained for. He was hungry all the time. At first this was understandable, for he had a hard time locating those foods that he knew to be safe to eat— onion and mariposa bulbs, cattail rhizomes, pine nuts and manzanita berries—and a hard time catching trout. Even later, though, when Mark was able to gather as much food as he could eat, he was still hungry. The artist, who describes his normal diet as a healthy diet of "mostly fish and not a lot of meat," began to crave foods that he never really thought he liked—foods like the hot school lunches of his past, lasagna, and "those thin little hamburgers with wilted pickles on top."

When I spoke with Mark soon after his return to civilization, he said that he had been craving those foods because they were familiar to him (unlike road-killed snake with cattail rhizome paste, one of his most memorable meals out in the wild) and because they were associated with a time in his life when he was being taken care of by others. But his cravings also had all the trapping of meat hunger.

Meat hunger appears often in ethnographic accounts, but it is not the only hunger experienced by human populations. It is not the only kind of community craving. Nor should we expect that it would be, given the constraints of protein metabolism and our physiological need for a steady supply of glucose. Meat hunger may be the most common type of hunger because animal foods are scarcer than plant foods at most latitudes. But if Mark Brest van Kempen had been camped on Coats Island with Josiah's family or if he had tried his experiment in the woods of Westchester, where there are plenty of deer to be eaten, but where deer have pretty much polished off all the edible plants, he would have experienced very different cravings. Then his cravings might have been for hot rolls and mashed potatoes, gum balls and jawbreakers.

The Mbuti, hunter-gatherers who live in the forests of Zaire, experience such cravings for carbohydrates and sugars, and during the honey season, they have a passion for honey, says anthropologist Colin Turnbull, that cannot be satisfied by any amount of alternative foods. Where they live, the Mbuti can find meat much more easily than high-energy roots and tubers. In fact, they probably would not be able to survive in the forests were it not that their techniques of communal net hunting are so successful that they can trade meat with nearby agriculturists for foods that are high in carbohydrates. The Mbuti keep a portion of each animal for themselves—the head, neck, and all the inner organs (just what one would expect of an energy-starved people)—but they swap the large cuts for vegetables, root crops, and cassava flour, their most favored trade commodity. Sometimes they even refer to their net hunt *kuya* as "a hunt for cassava flour." Half a kilogram of meat gets them a kilogram of flour, a trade—like that of caribou for seaweed in the Arctic—that only a large-brained omnivore would make.

Twelve

THE NATURE OF FOOD

In the years before people like Jane Goodall, Katharine Milton, and Ken Glander took to the savannas and forests of Africa and Central America to observe troops of monkeys and apes in the wild, there were any number of convincing, seemingly absolute criteria that separated Homo sapiens from the rest of the animal world.

Charles Darwin thought it was tools and the use of tools that started the cascade of human evolution. When our early human ancestors first began to walk on two legs, the great theorist conjectured, their arms and hands were freed to develop and use tools. But little did Darwin imagine, in those days when so few animals had been closely observed in their natural habitats, how many different animals use tools. Crows use cars as nutcrackers, and they hold twigs in their beaks to pry for insects. The woodpecker finch of the Galápagos, one of the birds that so informed Darwin's thinking, digs out grubs with a cactus spine. How could Darwin know that chimpanzees "fish" for termites and "dip" for ants using long twigs or blades of grass that they insert into the nests of their prey, as Goodall first observed in 1963 in Gombe National Park? Or that chimpanzees living in the rain forests of the Ivory Coast use stones to break open the hard-shelled nuts of *Panda oleosa* trees, a phenomenon reported only as recently as the early 1980s by biologists Christophe Boesch and Hedwige Boesch-Achermann? Many of the tools that humans use are more complicated and more diverse than the tools used by other animals, but clearly tools cannot be the Rubicon of human existence.

For other theorists, it was hunting and a reliance on meat as a protein source that differentiated humans from other animals. Many animals hunt, of course, and social carnivores like lions, hyenas, and African hunting dogs hunt cooperatively to bring down prey much larger than themselves, but these animals are specialized carnivores with specialized digestive systems and specialized teeth for killing prey and ripping apart flesh. Humans have none of these adaptations, and yet humans, in almost every part of the world, eat a great deal of meat. Might not it be this aspect of our diet and behavior that was responsible for all the other changes in humans, including our large brains and complex material culture?

"Hunting is the master plan of the human species," University of Wisconsin anthropologist William Laughlin wrote in 1968; it is the organizing activity that integrates the morphological, physiological, genetic, and intellectual aspects of individual human organisms. Laughlin made the most unequivocal expression of the "hunting hypothesis" in the scientific literature, but by 1968 this hypothesis was the prevailing doctrine. It got its start in the 1920s with Raymond Dart's discovery of the first Australopithecus skull and his belief (now judged to be mistaken) that Australopithecus, the first bipedal primate, was largely a meat-eater and was responsible for the many stone tools near the site where it was found and was later popularized by Robert Ardrey in *African Genesis*. In the scenario of Laughlin, Dart, Ardrey, and many others, Australopithecus was the first primate to use tools, and the first tools that Australopithecus used were weapons for bringing down prey and butchering tools for skinning and cutting up large carcasses.

This is the just-so story that most of us grew up on, and it's certainly easy to see why it was so convincing for so long, why scientists and laypeople alike have had a hard time letting it go. Australopithecus may not have been much of a hunter (those stone tools that Dart found in association with the Taung child have since been attributed to contemporaneous populations of Homo habilis and/or Homo erectus), but one of the most conspicuous things about most human populations *is* their hunting skills, their ability to kill and process large animals. Early sites that *are* definitely linked with the Homo line are littered with bones and stone tools.

But it's also easy to poke holes in the many false assumptions upon which this hypothesis rests, primarily the assumption that hunting was a hominid innovation and that the rest of the primates were and are vegetarians. "We are vegetarians turned carnivores," while other primates are "typical forest-dwelling fruit pickers," asserted Desmond Morris in 1967 in his best-selling *The Naked Ape*. Several years later, even two such prominent scientists as George Schaller and Gordon Lowther reiterated this view. "All monkeys and apes were basically vegetarians feeding on leaves, fruits, roots, and bark," they wrote, even though Jane Goodall had already published detailed accounts of chimpanzees in Gombe spending long hours "fishing" for termites and ants, of chimpanzees hunting for monkeys, peccaries, and antelopes.

Perhaps it was because Goodall was not trained as a scientist or because her observations were forcing people to rethink their favorite theory of human evolution, but her reports of chimpanzee hunting were greeted with much skepticism by the scientific community. They were dismissed as deviant behavior on the part of the chimpanzees, behavior that might have been caused by the fact that Goodall had provided them with bananas in order to habituate them to her presence. She was widely criticized for this technique, and it could well have had an effect on the chimps' interactions and behaviors.

But later studies in the same, and different, populations of chimpanzees substantiated Goodall's first claims, and scientists gradually came to accept the fact that humans are not the only primates to hunt for meat. Even baboons hunt, as revealed by anthropologist Shirley Strum in her seven-year study of the predatory behavior of this species of African monkey. Some scientists tried to keep the hunting hypothesis alive by asserting that hunting is qualitatively different in humans because only humans hunt cooperatively . . . or because only humans consume a large quantity of animal flesh . . . or because only humans willingly share the meat that they kill. But more recent studies have let the wind out of even these sails.

In the rain forests of the Ivory Coast, Boesch and Boesch-Achermann, a husband-and-wife team who spent five years habituating chimps to their presence before they could even begin their observations (five long years, during which they must have been sometimes tempted to use bananas or some other treat), found that chimpanzees hunt cooperatively—at least in this environment. Here chimpanzees move together in a highly coordinated fashion to trap monkeys and block off escape routes. Alone or in pairs, rain-forest chimps are successful in catching monkeys less than fifteen percent of the time. When three or four adult males work together, their success rate more than triples. These are odds that the chimpanzees seem well aware of, as Boesch and Boesch-Achermann point out. Of the two hundred monkey hunts they observed, ninety-two percent were cooperative affairs. Even baboons exhibit complex hunting behaviors involving more than one baboon, as Strum has reported. In Kenya's Great Rift Valley, they team up as relay runners, for instance, to outrun an antelope.

And just recently Craig Stanford's study of hunting practices in Gombe finished off the argument that humans are different because they are the only primates that consume large quantities of meat. This anthropologist from the University of Southern California saw chimps hunting with such gusto that they laid to waste one-fifth of their territory's population of red colobus monkeys each year. Chimps, he found, consume up to a quarter-pound of meat a day when they hit their hunting stride, an amount that equals the daily amount of meat consumed by some contemporary hunter-gatherers and that far exceeds that consumed by individuals in many agricultural societies. Chimpanzees are still primarily vegetarians, Stanford notes, but meat-eating is both a persistent and nutritionally significant activity. "Chimps absolutely love meat and get extremely excited about hunting," comments Richard Wrangham, who has seen evidence of chimpanzees preying on at least twenty-five different species of mammals. "They will wait for an hour under a tree for just three drops of blood to fall off a leaf." And why not? Chimps, like humans, are omnivores with large brains that have gambled their existence on finding high-quality foods.

So real-life observations of chimpanzees and baboons showed that neither tools nor hunting was the thing that turned apes and monkeys into humans. But they did give scientists much to ponder. For instance, before these studies of chimpanzees and baboons were published and gradually accepted, who would have ever thought that apes and monkeys had anything to tell us about the origins of the human sexual division of labor and the fact that men hunt and women gather in the great majority of hunter-gatherer societies? But chimps also exhibit sex differences in their foraging behavior, these scientists observed. And while these differences do not represent a true division of labor as it is practiced by all human societies (a division in which men and women collect and willingly share two very different kinds of foods), they could be the melody on which hominid populations harmonized as they confronted the changing environments of eras long ago.

All of a sudden what seemed important was not how animals like chimpanzees and baboons differed from humans but how similar they were in important ways. The behavior of these animals in their natural habitats does not suggest an abrupt and absolute break between the behavior of humans and other primates (a break that in the end would be difficult to explain given that we share some 98.4 percent of our genes with chimpanzees) but a subtle shift. "Selective forces were not operating on a tabula rasa [when hominids began their evolution]," as Shirley Strum puts it, "but on a primate species which already had certain predispositions." Contemporary populations of chimpanzees and baboons are not, of course, a precise analogue of those earlier evolving populations, but they may be a way of understanding what some of those predispositions were.

Sexual differences in feeding strategies are not uncommon, as we know. They originate in different species for different reasons: as a form of cooperation; as a way that males can get closer to females without driving the females away; and/or as a result of the different nutritional needs of males and females. In chimps, though, the different ways that males and females feed themselves seem tantalizingly familiar, a slightly altered version of the way that human males and females do things.

In a study of the same troop of chimpanzees that was first
habituated by Jane Goodall, William McGrew of Miami University in Ohio found that female chimpanzees spend three times as
much time gathering termites as adult male chimpanzees, but male
chimpanzees were the only ones to hunt for baboons, peccaries,
and monkeys. McGrew observed thirty predatory incidents during
the year of his study, and in none of these did females take part in
any killing. Subsequent studies have shown that female chimpanzees can and do hunt, but they hunt much less than males.
Geza Teleki, a primatologist at George Washington University,
also observed the hunting practices of chimpanzees at Gombe
and reported that some kills—less than four percent of those
observed—were made by adult females. Japanese primatologist
Toshisada Nishida reported that females were the hunters in nine
out of the thirty-three instances of predation that he observed at
Kasoje in the Mahale Mountains. So by and large, the pattern, first
noted by McGrew, has stood. Certainly, no one has ever seen its
reverse: males concentrating on insects and females on meat.

So how do scientists explain this intriguing pattern? For
McGrew, the different techniques necessary to capture insects and
mammals seem to be the key. "Generally," he explains, "Gombe
chimpanzees capture their primate prey high in the trees; acute balance, sudden direction change, concentrated bursts of exertion,
and manual dexterity are very important. The activity is erratic,
irregular, and hurried. Transporting an offspring, pre- or post-
natally, would hamper all of these.

"In contrast," McGrew continues, "social insects [termites
and ants] represent a stable, localized animal food source perhaps
more suited to female species-specific characteristics. Once at the
mound, the chimpanzee's termite-extracting process is sedentary

and interruptible. The fishing technique requires little more than forelimb motion, and young infants may cling ventrally and sleep, suckle, or watch while the mother fishes."

Sedentary and interruptible. Another scientist described the act of termite fishing as "self-paced and interruptible," and when I read those words, I felt that I was reading about myself and the way I organize my days, or did when my children were younger and living at home. As I remember it, nearly everything about my life had that self-paced and interruptible quality. It's not that I didn't like to throw myself completely into a project. It's not that I wasn't able to become totally immersed in what I was doing. But there was always a part of me that was ready to switch gears—and change a diaper, make an after-school snack, or discuss a difficult math teacher. There was always a part of me that I kept available, in reserve. This was an anathema, I know, to my husband, who likes to work at what he is doing until he falls over, but most of my women friends understand it well.

Those female chimpanzees that do hunt at Gombe and at Kasoje help us to understand why most females don't hunt and the constraints that most chimpanzee females face. One of the adult females at Gombe is sterile, and she behaves much more like a male in many ways. She roams more widely than the other females and attends many more sessions where males have killed prey and are dividing up the carcass. At Kasoje the females that hunt are much more likely to kill a young antelope or gazelle than a monkey since antelope young can be seized from the hiding spots where their mothers have left them, but monkeys have to be chased through the trees. Evidently, it is not the taste for meat that keeps most females from hunting, but the tactics involved, tactics that might jeopardize the care and safety of primate young. If a mother was to hurt herself during a hunt, she would threaten both her own survival and that of her offspring, for primate young are dependent on their mothers for much longer than other mammalian young.

In her study of predatory behavior in baboons, Shirley Strum also found that female baboons are much less likely to hunt than males. And when they do hunt, they are much more likely to pursue small game, such as hares and birds, than adult antelopes or gazelles. Strum, too, attributes these differences in hunting behavior to the problems of infant care since, as she has found, the female baboon's interest in meat is great and since she is as physically capable as males.

Some may not like the implications of these sex-linked differences in behavior. Some may find their familiarity uncomfortable rather than revealing and eagerly point to the few examples of societies in which women do hunt—the Agta of the Philippines, for example, or the Mbuti of Zaire—as proof that the roles usually assumed by women are imposed by social conventions rather than biological constraints. But far from being examples of how societies used to exist before labor was artificially divided between men and women, the Agta and the Mbuti seem to be exceptions that test, and prove, the rule. In both, nursing women hunt less than women who are not nursing, and in both, meat is traded for the garden products of neighboring agriculturists. The Mbuti and the Agta are forest dwellers, and most tropical forests, as anthropologist Thomas Headland was one of the first to point out, do not contain enough naturally occurring, carbohydrate-dense tubers to fully sustain a human population. The tropical forest was home to our primate ancestors, but by the time hominids had acquired their large brains, they couldn't go home again. At least not without some way of obtaining adequate amounts of energy-rich foods. Agta and Mbuti women hunt, most probably, because there is little for them to gather. They hunt so they can trade their quarry for the carbohydrates that their families need to be properly nourished.

"Females trade off alternatives that give different benefits to their children," Kristen Hawkes tells us in an article in The Archeology of Human Ancestry. In the case of the Agta and the Mbuti, women trade a little of their children's safety for the ability to survive in tropical forests. In other groups of hunter-gatherers, women specifically avoid taking resources that would increase health risks to their young. Among the Ache of eastern Paraguay, for instance, women collect the honey of stingless bees; but only men go after the hives of bees that sting.

The role of the hunter actively pursuing its prey may not be the role that most primate females play, but that doesn't mean that primate females are without significant roles in life. Most primate mothers give their young all the care and the food they need to grow to independence. Primate males often help with group defense, but very few care for their young directly. In many hunter-gatherer societies, the foods that women pursue, plants and small animals, provide their families with the majority of their calories. Among the aborigines of Australia, women value men's occasional contribution of large, red-meat animals, but they do not depend on that contribution for their survival. Aboriginal men, on the other hand, do depend on the foods their wives provide. Among chimpanzees, females are the technical wizards of the society and use tools much more often than males, a startling finding that may have important implications for the real role of tools in human evolution.

Early theories about human origins, as we know, closely linked tool use and hunting. Tools enabled early humans to hunt, it was thought, and the first tools were used in hunting and butchering.

That was before we had firsthand knowledge of what animals like chimpanzees do when left to their own devices in their native habitats. But who uses tools the most in the chimpanzee populations that have been closely observed? And for what reason? Male chimpanzees, it turns out, are successful hunters without the aid of any tools. Male chimps have been known to throw an occasional rock at a potential prey (or a rival), but hunting chimp-style doesn't require any special equipment. Cunning, speed, agility, yes. But no equipment. Female chimpanzees, on the other hand, use tools routinely to gather and process their foods. Females fish for termites and dip for ants much more often than males, and each fishing and dipping session requires a tool made of a strip of bark or a blade of grass, modified somewhat by shortening, narrowing, or stripping it of leaves.

In the rain forests of the Ivory Coast, the only place where chimpanzees have been seen using stone hammers to crack open nuts, females also use these tools the most. They crack more nuts than males, as we know, and they crack nuts when they are dry or fresh, on the ground or in the trees. Males and females also differ here in that only females appear to plan their nut-cracking episodes. Nut-cracking requires both an anvil—the hard surface of a stone or a root—and a hammerstone. But hammerstones are hard to find in the forest, and would-be nut-crackers are well advised to keep track of their implements and move them about from tree to tree. Male chimpanzees have been found to crack nuts opportunistically—that is, they crack them when they see a tree with nuts and a stone with which to open them. But female chimps plan ahead. Females are the only ones that have been seen transporting a stone from nut tree to nut tree.

And so it is in every study of chimpanzees that has looked at sex differences in behavior: females use tools much more often than males. This has led some scientists to the conclusion that human technological skills probably did not originate with hunting but with the collecting of plants, nuts, and insects. "The grasses, vines and sticks used by chimpanzees for so many purposes," muses Geza Teleki, "are excellent prototypes for the digging sticks carried by many hunter gatherers." These sticks, which are used almost universally to dig out roots and tubers and poke for insects and small mammals, are associated with women and women's work in many societies. And again, why should we be surprised except for the fact that hunting, tools, and male behavior have been linked for so long in our minds? Most females have a harder time meeting their energy needs than males. That is the nature of being female. So why wouldn't primate females be the first to develop tools to help them in this quest? (It would be interesting to know whether animals other than primates also show sex differences in tool use— whether the female sea otter, for example, is more likely to crack conch shells on her stomach—but as far as I know, no studies have addressed this question.)

At some point in human evolution, it is clear, stones were sharpened into axes and cleavers, and axes and cleavers were used for hunting and butchering large prey. But observations of tool use in chimpanzees give us a new way of looking at the evolution of tools, and a new explanation for the fact that the earliest *recogniz-able* hominid tools—stone tools found in association with the fossilized bone of prey animals—date back only two and a half million years, while hominid origins go back four and a half million. Because stones and bones endure while parts of plants do not, the archaeological record has been biased toward tools that were used in hunting and butchering. But those tools were no doubt late-comers in the human tool chest. They were descendants of much earlier generations of tools—tools that, like the chimpanzee mother's blade of grass, would have disappeared without a trace. The *first* hominid tools may have been unprocessed stones used for pulverizing roots, crushing seeds, opening nuts, and smashing the long bones of hunted or scavenged prey in order to extract their contents of fatty marrow. They may have been long, sharpened sticks for digging roots and tubers, ostrich and tortoise shells for carrying water, and animal skins for transporting meat or babies.

My husband, the toolmaker, doesn't quite buy this scenario of human technology having its roots in female tool use. Tools are his life, and they are my nemesis. The only tools with which I can be said to have a good relationship are my gardening tools, my weeding swoe, and my cart—simple, sturdy tools. Not every woman is this way, I realize; not every man is as handy as my husband. Just many of the women and some of the men I happen to know. But the scenario I am envisioning has little to do with technology as it came to occupy humans, the technology of today or the technology of the stone age. Rather, it deals exclusively with the technology of the earliest humans, with technology and tool use when women foraged for themselves and raised their children alone, as chimpanzee mothers do, before their mates helped with this arduous task.

Once human males bonded with individual human females, they too would have had reasons for developing their skills with tools (to better provide for and defend their families), reasons for taking on tasks that get in the way of child care. Women, meanwhile, would continue to develop their own skills, skills that would tend to be "self-paced and interruptible" but every bit as important to the survival of their children. Men are better at throwing things than women, at map reading and navigation, and at mentally rotating objects to see how they fit together, but women are more verbal, observant, and meticulous. They tend to be better than men at identifying matching items and are faster at certain manual tasks involving precision. "There is," as writer Matt Ridley points out, "abundant material for those who like stereotypes here, but none of it says anything about the woman's place in the home." For as long as the pair bond has been a part of hominid life, both men and women have gone out to work: one to hunt, the other to gather.

Nicholas Roth, an experimental archaeologist who has spent many years re-creating prehistoric flaked-stone technology in order to understand how our distant ancestors made and used tools, offers an interesting piece of circumstantial evidence that tool-making in humans originated with females. When Roth struck core stones with other stones in order to make sharpened stone flakes of the kind that Homo habilis and Homo erectus used for killing and butchering animals, he noticed that his flakes had a certain pattern or orientation to them because of the fact that he was right-handed. The majority of his flakes had a right as opposed to a left orientation, and when Roth later examined flakes that had been dug from the Koobi Fora site in northern Kenya, a site with archaeological remains up to 1.9 million years old, he determined that those flakes had the same orientation as the flakes he had made. He concluded that right-handedness was a characteristic of the hominid line as far back as 1.4 to 1.9 million years.

Roth discusses some of the important ramifications of right-handedness for human brain development, including the fact that the left cerebral hemisphere, which is dominant in right-handed individuals, controls language, time-sequencing skills, and the ability to conceptualize the future. But he doesn't address the question of why humans became right-handed, a characteristic unique to our species. Other animals have individual preferences and favor one leg or one paw over another, but those preferences even out over the course of a population, and an animal is no more likely to be right-legged or left-pawed than one would expect on the basis of chance alone. Over ninety percent of humans, though, are right-handed, with the percentage somewhat higher in women than men. But why would right-handedness have happened, and why would women be more likely to be right-handed than men? What difference does it make whether you use a tool with the right hand or the left? (I'm asking, of course, about a primitive tool and not one modified for right-handers.)

Many anthropologists today make the assumption that right-handedness is simply a by-product of the fact that the left cerebral hemisphere did become dominant in our species, that it is an aftermath of the necessity for increasingly subtle forms of communication in hominid social interactions—i.e., language. But what if we consider the possibility that right-handedness came first and was the development that enabled those increasingly subtle forms of communication, the development that allowed our ancestors to speak? Then we do need an answer to the question of what difference it makes if you use a tool with the right hand or the left. And the answer is "None whatsoever"—unless you are carrying something in your left arm. And what do humans carry in their left arms much more often than their right?

Babies.

The pediatrician Lee Salk was one of the first to note that mothers tend to carry their infants in their left arms. It was a behavior he attributed to the soothing effect of the maternal heartbeat. Subsequently, other pediatricians have shown that human infants have an innate preference for turning their heads to the right. And they have observed that mothers who hold their infants in their left arms can make more eye contact with their babies and observe them for any signs of distress, an early detection system that must have been as important to the first hominid mothers as it is to mothers today.

Or perhaps even more important, for holding babies was a brand-new kind of mothering back then, necessitated by the fact that hominids walk on two legs.

The shift to bipedality required hominids to develop a narrow, rigid pelvis (one that can bear more than half the weight of the skeleton), but a narrow rigid pelvis means that babies, especially full-term babies, have a hard time passing through the birth canal. The solution to this very real impasse or bottleneck was that hominid mothers began giving birth at an earlier and earlier stage of the fetus's development, and they compensated for their infant's prematurity with increasing amounts of care. Chimpanzee infants must be partially supported for the first few weeks of life, but then they are strong enough and coordinated enough to move around on their own and to cling to their mothers entirely by themselves. But for two years or so, human infants—born some six months before the time in utero that they would need to be as advanced as ape infants at birth with regard to motor skills and brain development—must be carried almost everywhere they go. They can't even help their mothers out by holding on to them since, in another adaptation to life out on the hot, open savannas, mothers had lost most of their body hair.

I think you can see the picture that I am trying to paint here: a picture of early human mothers struggling to care for these premature infants *and* find enough food to survive. A picture of mothers holding their infants in their left arms as they dig, probe, and smash nuts and bones, using right-handed techniques that their children will copy as they grow. A picture, too, of these right-handed mothers and their children becoming better at both communicating *and* finding food since, as Steven Pinker points out in *How the Mind Works*, the left hemisphere of the brain is the seat not only of language but of being able to recognize and imagine whole objects by their parts. The seat, in other words, of being able to think of a plant when all one sees is a seed, of imagining a buried tuber when all one sees is a withered stalk.

But how did hominid females come to also enlist the help of their mates? How did the pair bond ever become a part of hominid life? Jane Goodall's early observations at Gombe have some bearing on this question, too, for not only did she see chimpanzees hunting and chimpanzees using tools. She was also the first to see chimpanzees sharing food with each other. Like hunting and tool use, food sharing was another one of those things upon which scientists had pegged their theories of human evolution, and Goodall's initial reports of this behavior were also speedily dismissed. In 1978 Glynn Isaac published a very influential paper in *Scientific American* in which he argued that food sharing was the innovation that made our ancestors human. He characterized sharing in troops of chimps as nothing more than "tolerated scrounging."

But tolerated scrounging just doesn't do justice to Goodall's description of a chimp requesting food from another animal: of how it peers intently into the face of the other chimp and reaches out to touch the food or the chimp's chin and lips; of how it softly whimpers and hoos. Food is not shared in a troop of chimpanzees as widely, or regularly, or willingly as it is in a band of hunter-gatherers. But just as the particulars of tool use in chimpanzees have made us rethink the beginnings of human technology, so the particulars of chimpanzees' food sharing—when, where, and how they do it—may tell us something about the origins of human food sharing. And that may tell us about the origins of the human pair bond.

I agree with Isaac about the importance of food sharing in human evolution, but it was not food sharing per se that gave this behavior its evolutionary power. For food sharing is, in fact, somewhat common in the animal world. Members of a pack of African hunting dogs bring food (in their stomachs) back to their den for a nursing mother; a pride of lionesses "share" the antelope that they have worked together to kill. (Though "tolerated sharing" might be a better name for *their* rough-and-tumble dinners.) In any number of species, males provide food for their mates and/or their young. Rather, it was the human twist on food sharing that makes food sharing in humans so different—regular sharing between males and females *and* females and males of two very different kinds of foods. In lions, males, which are substantially larger than females, often usurp the carcass of a kill that the pride had made. But no animal shares with us this one simple behavior: males and females willingly providing food for their children *and* their mates; males and females willingly sharing with each other and their offspring.

Most of the sharing of food among chimpanzees, as in the rest of the animal world, takes place between mothers and their dependent offspring. Mothers are the primary caretakers in the great majority of animal species, and those who give their offspring food, whether it be milk or solid food, give them a head start in this world. Mothers that continue to provide food for their young after weaning help them to survive the difficult transition to independence, the period of highest mortality for many young animals. This behavior makes so much sense that most of us take it entirely for granted. It seems so logical and familiar that we tend to forget that there are mothers out there who don't ever feed their young and that feeding, like all forms of parental care, has its costs.

So even this kind of self-evident behavior can have dynamics that are revealing. Chimp offspring learn to feed themselves as they are being weaned, and in most parts of Africa where chimps are found, weaned youngsters feed themselves entirely. This is true at Gombe, though mothers occasionally allow offspring to remove termites from their "fishing" poles. But in the rain forest of the Ivory Coast, weaned offspring still get much of their food from their mothers. There, mothers have been found to share some sixty percent of the panda nuts that they crack with their young. They continue to share with them until those young are old enough— six years or so—to open those difficult-to-crack nuts themselves. Mothers in the rain forest also share other foods that they acquire with tools—honey, ants, and bone marrow. The difference between these mothers and the mothers at Gombe, Boesch and Boesch-Achermann propose, lies in the environments in which these two populations of chimpanzees live: the rain forest of the Ivory Coast versus the savanna-type environment of Gombe National Park. Life in the dense, dim forest, these researchers say, may be more difficult than we ever imagined—at least for the chimpanzee. Foods may be harder to find and more difficult to process. Just as chimp males need to hunt cooperatively in the rain forest in order to have any hunting success, chimp mothers may need to share in order to ensure the survival of their young.

Back at Gombe, Jane Goodall also observed chimpanzees sharing food outside the mother-infant dyad, and it was this behavior that really surprised the world and that anthropologists were quick to dismiss. It was this behavior that humans would prefer to keep as their own exclusive province. Food sharing of this kind happens less often than food sharing between a chimpanzee mother and her offspring, but the dynamics of these exchanges are even more revealing.

First of all, the food that is most often shared among chimps, other than mothers and their offspring, is meat, the carcass of an animal after a kill has taken place. That is the food that is the most often fought over, stolen, begged for, and shared. In Gombe, after a male chimp has successfully cornered and killed a colobus monkey, all the males in the troop spend hours and hours squabbling over the carcass and feasting on the meat. From the amount of interest that chimpanzees have in fresh meat, one suspects that they too experience meat hunger, that they too might be dreaming of monkey liver and thin colobus strips.

Meat plays an important role in chimpanzee societies for probably the same reasons that it does in human societies. One, it contains fats, minerals, vitamins, and all the essential amino acids in all the right proportions for body building and optimal body function. Two, protein, at least in humans, has been found to have a greater effect on satiety, or the suppression of appetite, than either fats or carbohydrates. If this holds true for animals other than humans, those that eat optimal amounts of protein would have more time to rest, or nurse their young, or assert their dominance, or woo females, before they again have to respond to the demands of an empty stomach. Three, meat is hard to acquire and it comes in a large enough package, so to speak, to make sharing possible—even advantageous. Excess amino acids cannot be stored in the body, and their metabolism actually costs the body calories and muscle mass. After a certain point, therefore, after an individual's protein needs have been met, lean meat (which most wild meat is) does a body no good. It is useless, even harmful, from a nutritional point of view. Moreover, it cannot be stored for the next day because at most latitudes, meat goes bad fairly quickly.

So what do primate hunters do with their excess, worthless meat? They could abandon it to those whose protein needs have not yet been met. Or they could use it to win females and influence friends. In the rain forest of the Ivory Coast, where chimpanzees hunt cooperatively and meat sharing is very widespread, the male who winds up in possession of a kill shares it with all the other chimpanzees who participated in the hunt. They in turn share it with their relatives and friends, very much as hunter-gatherers in Africa or the Arctic divide up the carcass of a zebra or whale. And like some human hunters who distribute the meat of their kill but reserve very little for themselves, chimpanzee hunters don't always partake of the food themselves. At Kasoje in Tanzania, Toshisada Nishida twice saw a dominant alpha chimp secure a big carcass that he scarcely ate but held for others to nibble on. "Who has but once dined his friends, has tasted what it is to be Caesar," Herman Melville says in *Moby-Dick*, and as this chimp also seems to understand. "It is a witchery of social czarship."

At Gombe, when a male chimpanzee shares a carcass with other males, the recipients of his generosity are not always the most dominant members of the troop, as one might expect. They are often the males with whom he is most closely related, or with whom he has a long-standing relationship of reciprocity (males who back him up in a fight and whom he backs up in turn), or the oldest and most experienced hunters in the troop (males from whom he might expect meat at some time in the future).

Males are not the only chimpanzees to consume meat. Females may hunt much less frequently than males in every location where wild chimpanzees have been observed, but their interest in meat is great and their attendance at kills is high. There they join males in begging for food, and there they too are not rewarded indiscriminately. Rather, as Geza Teleki was the first to observe in the 1960s, males in possession of a kill give meat preferentially to females that are sexually receptive, females that are in estrus. Craig Stanford, who recently spent four field seasons at Gombe documenting the chimpanzees' hunting tactics, has substantiated Teleki's observation. Stanford found that the single best predictor of when Gombe chimps will decide to pursue a group of colobus monkeys (instead of allowing them to feed in peace) is the presence of one or more swollen, sexually receptive females in the party. Time after time Stanford saw male chimpanzees dangle a dead red colobus monkey in front of a swollen female, sharing it with her only after she allowed him to mate.*

"Since hunts also occur when no estrous females are present, this trade of sex for meat cannot be the exclusive explanation [for hunting]," observes Stanford, "but the implications are nonetheless intriguing . . . chimps use meat not only for nutrition; they also share it with their allies and withhold it from their rivals. Meat is thus a social, political, and even reproductive tool."

---

*Male bonobos or pygmy chimpanzees, a distinct species of chimpanzee that lives in the forest and was only discovered as recently as 1929, have also been observed to trade sex with a female in estrus for desirable foods. But unlike savanna chimpanzees, bonobos also use sex—between males and females and females and females—to diminish the daily stress of competition over food. For bonobos in zoos and in the wild, casual sex, sex not necessarily associated with reproduction, seems to make it possible for them to eat in peace.

After observing the bonobo colony at the San Diego Zoo, Frans B. M. de Waal, professor of psychology at Emory University, remarked, "As soon as a caretaker approached the enclosure with food, the males would develop erections. Even before the food was thrown in the area, the bonobos would be inviting each other for sex: males would invite females, and females would invite males and other females." From Zaire's Lomake Forest, Nancy Thompson-Handler, of the State University of New York at Stony Brook, described seeing bonobos engage in a flurry of sexual contacts lasting for five to ten minutes when they entered trees loaded with ripe figs or when one of them had captured a prey animal. Only then did they settle down to consume the food.

Among baboons, the only food other than milk that any baboons have ever been seen to share, even mothers and their offspring, is meat. And baboon interactions over this substance have many of the same overtones that they do in chimpanzees. In the troops studied by Shirley Strum, males don't actually give meat to females in estrus, but they do sometimes allow females that they are consorting with to approach them while they are eating meat and to take up eating beside them. Female baboons also take advantage of the conflicts between males to steal a temporarily neglected carcass—and they even invent conflicts to achieve these ends.

Strum described the techniques of one female: "First she tried to groom the male into a state of extreme relaxation and obtain some meat while he was not paying attention to her. The male soon realized the effectiveness of this tactic, however, and continued to allow her to groom him, but prevented her from actually seizing any meat. Her next tactic was to direct aggression at a female with whom the male was in association, thus creating for the male a situation of conflict between the meat and his sociosexual interests. The male then abandoned the meat and the female aggressor obtained it." Baboon males, as Strum has found, always place their sexual and social interests above meat-eating. Baboon females, on the other hand, may be more interested in meat-eating than in any other activity.

Now this is uncomfortable, the idea that here is where it might have all begun, life as we know it, here in this trade of meat for friendship and meat for sex. But it does have the ring of truth about it, for who would argue that there is not an implied element of trade in the dinner date, where an expensive meal is often the prelude to intimacy? Or the dinner party, where guests are usually expected to reciprocate at some future time? It makes sense of the central role that meat plays in most societies, in feasting, sacrifice, and everyday meals. And it makes sense given the nutritional value of meat and the fact that meat is a scarce resource at all but the highest latitudes. It makes sense given what it means to be a male or a female and the nutritional needs of females. Some of the peculiarities of hunter-gatherer societies can be understood only by peering through the lens of this exchange. In promiscuous societies, like those of the Ache or the Hadza, where men frequently have children by more than one woman, men who are the best hunters are most often named by women as their sexual partners. And they spend much more time hunting than men in societies that are largely monogamous, the !Kung or the Hiwi. Why knock yourself out hunting, these monogamous men seem to be asking, when you can only have one wife and can only sire just so many kids?

But human males hunt for meat not just to mate with more women but to give to their own children, an important difference between humans and chimpanzees. At some point in human evolution, as Matt Ridley puts it, male hunting changed from being just a seduction device to being part of a deal with one's wife. At some point, women began to provide food not just to their children but to their bonded mates, and groups of humans began to share food widely and extensively. Trading meat for sex and friendship may be where it all began, but that is not where it ended, at least in humans, the species with the most brain capacity and the most capacity for compassion and cooperation. We may never be able to discover exactly how, when, and why humans put their own human twist on food sharing (though in the next chapter, I will try), but each new study on chimpanzee and baboon behavior gives us a better understanding of where we started from—a better understanding of the nature of food.

Thirteen

THE HUMAN NATURE OF FOOD

There is something I've never gotten used to on these trips to the Arctic and to Central America that I've been lucky enough to take, something akin to wearing my clothes inside out. And that is living in a group of people, ornithologists or primatologists, who do so little for each other during the course of the day, people who cook most of their own meals and who don't often share either their food or their labor.

As a wife and a mother, someone who is used to living in a tight-knit family group and used to cooking and sharing two or three meals a day, it is oddly unsettling to step out of my roles of cook and caregiver. I always knew that food brings people together, but until those visits to faraway places, I never knew how uncomfortable I was in situations where the people I was with weren't sharing the same foods. You might think that I would have been glad for the break from my routine, glad to stop cooking and worrying about whether everyone had what they wanted to eat.

But before I had been away long, I would find myself making soups and baking cookies and cobblers, showing my regard and appreciation for these new people in old familiar ways. *You can take the mother out of the home*, I thought, *but you can't* . . .

Eat you alone; hungry you alone, they say in the Virgin Islands.

He who eats alone, chokes alone, is an Arab proverb.

"The idea of eating alone and not sharing is shocking to the !Kung," social anthropologist Lorna Marshall once remarked. "It makes them shriek with an uneasy laughter. Lions could do that, they say, not men."

"Taking food alone tends to make one hard and coarse," wrote the German philosopher Walter Benjamin. ". . . For it is only in company that eating is done justice; food must be divided and distributed if it is to be well received."

Most animals feed and forage for themselves, but humans share food every day. Sharing food is an integral part of our relationship with others. It is part of almost every human gathering, tragic or triumphant. For the ancient Greeks, a likable person, a sympathetic person, was someone with whom one could enjoy a symposium—dinner followed by much talk and drinking. Our companions, those we eat bread with, are our friends. In many cultures, two people do not feel they can talk in a friendly way unless they have eaten together. It is the equivalent of being properly introduced.

While chimpanzees share food more than was once thought, regular food sharing is not a characteristic of primates other than humans. When Canadian anthropologist Birutê Galdikas was studying orangutans in the forests of Borneo—and bringing up her own child at the same time—she noticed a sharp difference in attitudes toward food between young orangutans and her own young Homo sapiens. "Sharing food seemed to give Binti [her daughter] great pleasure," Galdikas observed. "In contrast, Princess, like any orang-utan, would beg, steal, and gobble food at every opportunity. Sharing food was not part of her orang-utan nature." Food sharing is not hardwired in the human psyche, as Turnbull's study of the Ik makes painfully clear (as few things are in a species like ours, which relies on flexibility and adaptability for survival), but it is part of us in all but the most dire circumstances. This reveals itself in many different ways.

At Gombe National Park, chimpanzees eat more when they are alone than when they are in a group. But humans eat more, much more, when they eat together. Based on seven-day food diaries kept by sixty-three adults, John and Elizabeth de Castro found that when meals were eaten in the presence of other people, increased amounts of calories were consumed. The meals were larger by as much as forty-four percent, and the participants in the study took in more of all the essential macro- and micronutrients. "It has long been theorized that social factors are a major influence on the eating behavior of humans," the authors write, and they conclude that in this age of overeating, eating alone may be "healthier eating."

Healthier in terms of calories, perhaps, but not so healthy if food sharing is important to our psychological sense of well-being.

In most but not all societies, some food is shared beyond the family unit. In all societies, food is an essential part of the relationship between men and women, a key to the universal custom of marriage. Husbands and wives may not always eat together. They may eat at different times or in different rooms or they may partake of different foods, but food is part of the intimacy between husbands and wives. In nearly all societies, husbands are expected to share their food (or resources) with their wife and children (or wives as the case may be); in *all* societies, wives are expected to share their food with their husband and children and to prepare food for their family almost every day. In countries in Africa, marriage for a man means being cooked for by a woman; for a woman, it means feeding a man. Cooking is often seen as the reciprocal of the coital acts of a man, and feeding a husband is such an important part of a wife's role that, in some societies, a wife is thought of as "the mother of her husband."

One cannot help but notice that a certain inequality exists in what is universally expected of the two sexes. Women are expected to sleep with and cook for just one man. Men may sleep with and be cooked for by more than one woman. Women direct all their resources toward their families. Men may direct some of theirs elsewhere. I will come back to this, but what is important here is that food, not just sex, binds human males and human females together, a fact of life widely reflected in the ceremonies and customs of different cultures. Weddings always include the sharing of food between friends, families, and most importantly, the bride and the groom. In some cultures the sharing of a certain special food is the entire marriage ceremony. In Western cultures, a wedding is not complete until the bride and the groom have shared the first piece of wedding cake.

Even in polygamous societies, customs reflect the link between sex and food. Cooking and sex usually take place on the same rotational basis so that the wife with whom the husband is sleeping is usually the wife who provides him with food. (But food is so important to marriage that a polygamist must at least sample all his wives' cooking.) An American polygamist and Mormon fundamentalist who was recently interviewed for the *New York Times* rotates his meals and nights between his three wives, but his entire family eats together on Sunday.

So how did we get to this point, where sleeping together and eating together are two sides of the same coin? Where, in parts of the world as far removed as Ghana and southern Brazil, there are verbs that mean both "to copulate" and "to eat"? Where, as M.F.K. Fisher wrote in the foreword to *The Gastronomical Me*, our basic needs, "for food and security and love, are so mixed and mingled and entwined that we cannot straightly think of one without the others"? When did women start feeding men, and when did men start feeling, like the !Kung hunter in John Marshall's 1968 film *The Hunters*, that "a wife with breasts full of milk for a new baby made him want to kill a fat buck?" A chimpanzee male loses interest in a female immediately after mating, but human males and females have close unions that can last for many decades.

If our earliest hominid ancestors resembled contemporary chimps and baboons in their social and sexual habits, how did we move from this "collector-predator existence," as one anthropologist has called it, to the existence of the hunter-gatherer, an existence based on stable bonds between males and females and on a sexual division of labor, with men and women acquiring and sharing very different foodstuffs?

Some are convinced that male hunting skills were the impetus for this novel arrangement of human beings, arguing that a consistent source of meat enabled males to share excess food with their mates and children. But primate bodies and brains don't run on animal foods alone, as we know, and although chimpanzees hunt regularly, you don't see male chimpanzees forming stable, long-term pair bonds.

Then there is that small but troubling inequality in what is expected of husbands and wives throughout the world. If hunting was the thing that allowed hominid males to begin caring for their mates and children, the thing that underlies the human pair bond and the human family, why are there societies in which men aren't expected to work as hard for their families as women? An early

description of aboriginal hunting and gathering practices in Australia noted that "women always took food back to the camp for
their husbands but if only a small quantity of animal food was
caught there was frequently none left for the women when they
returned." Why are there societies in which it is customary for
men to give their food to their sisters and their sisters' children—
and not to their wives and their own children—while there are no
societies in which women direct their resources anywhere but to
their own, immediate families? In tribal villages in Cameroon, men
don't consider it correct to provide food to their wives. Any food
or money that they acquire goes to their natal families.

Hunting must have played a significant role in the course of
human events, especially hunting in the grasslands that were re-
placing the forests around eight million years ago, when hominid
populations were beginning to diverge from ape populations. For
these grasslands provided hominids with many more hunting and
scavenging opportunities.* But still, knowing what we know about
females and their unending quest for food, knowing what we

---

*In the past decade or so, the role of scavenging in the early human diet has
received considerable attention, with several researchers suggesting that
scavenging was more common than hunting two million years ago. The same
nutritional constraints concerning protein metabolism apply regardless of
how one obtains one's meat, but a scavenged meal might well be a fattier
meal and therefore more suitable for hominid needs. It would be heavier on
bone marrow and brain tissue, tissues large carnivores would leave behind.
Doubtless, early humans depended on some combination of hunted and
scavenged animals to obtain their animal protein, as hunter-gatherers like the
Hazda do today. During our conversation in Arusha, the Hadza hunter
Abedi Shimba told me that there is no loss of pride associated with bringing
home scavenged meat. The Hadza do not hunt lions, he also said, because
lions provide them with food, though this might also have to do with the
fact that the livers of strict carnivores such as lions are often poisonous,
causing hypervitaminosis A.

know about when and why male chimpanzees share meat, wouldn't it be prudent to ask whether the gathering techniques necessitated by the savannas might also have something to do with the adoption of this new lifestyle?

Food in the savannas, for the fairly large-brained primates that our ancestors were, was more spread out, or dispersed, than it was in the forest. It would have included many energy- and nutrient-rich foods, among them berries, fruits, nuts, shoots, honey, shallow-growing roots and tubers, fruiting bodies of fungi, reptiles, eggs, nesting birds, some fish, mollusks, insects, small mammals, and the carcasses of animals that had been killed by lions and other carnivores. Most scientists agree that bipedality was an adaptation for finding these high-quality, widely dispersed foods and that walking on two legs allowed hominids to travel faster and farther in this new environment. Bipedality is, as Henry McHenry and Peter Rodman, paleontologists at the University of California at Davis, observe, "an ape's way of living where an ape could not live." With it, we became as good at covering long distances as our cousins are at climbing trees.

But bipedality, as we know, did not come without costs—especially for females. Bipedality made childbirth and child-rearing, jobs that were already difficult for the primate female, even more difficult. Mothers who found ways to lighten their loads would have been at a considerable advantage. Their children would have been more likely to survive and reproduce and to pass on those valuable child-care tips. Like some primates, and many women today, early hominid females probably turned to their sisters and mothers for a certain amount of help with their young.* Like chimpanzee females,

---

*The *grandmother hypothesis*, in fact, suggests that menopause, a reproduction feature unique to humans, came about because of the importance of

they probably invented tools to help them in this task, tools both for gathering foods and to help with child care.

We will probably never know when the first hominids started to use skins and woven fibers as carrying devices because these relatively fragile materials rot and disappear over time, but there are reasons for believing that the first hominids to do so were female. Necessity is the mother of invention, as we have so often heard, and necessity in this case was a hairless mother with a helpless hominid child. Also, carrying sacks are closely associated with women and women's work in many traditional societies. Among the !Kung, the *kaross*, an animal skin used for carrying, has so many feminine connotations that the knot—*!kebi*—that ties the *kaross* at the waist is also an affectionate term for "women." Richard Lee,

---

grandmothers to the survival of their grandchildren. This hypothesis, first proposed by George Williams in 1957 and publicized in recent years by the very charismatic anthropologist Kristen Hawkes, makes the point that it isn't the number of children you have that's important, in the great Darwinian scheme of things. It's the number of surviving grandchildren and great-grandchildren, great-nieces and -nephews. At some point, the argument goes, it makes more sense for women to stop reproducing long before they come to the end of their lives and to help care instead for their children's children.

There is some evidence to support this hypothesis (among the Hadza of northern Tanzania, older postmenopausal women spend significantly more time foraging and digging for roots and tubers than women in any other age category, and their activities have been correlated with the rate of weight gain in their grandchildren). But it has also been suggested that menopause evolved for a more direct reason: because human children require two decades or so of care in order to become self-sufficient adults, a much longer period of care than any other kind of offspring. In this scenario, menopause and the long postreproductive span of human females give females time to devote to their youngest children before they grow old and tired and near death. Either scenario, though, underscores the many difficulties that human mothers face in child-rearing.

one of the first anthropologists to recognize the importance of carrying devices in the lives of women, gives an example of how that term might be used. "For example," he writes, "if you hear the sound of voices in the bush heading toward camp and you ask, 'Who's that?' the response will often be, 'It's the !kebesi.' It's the knots." Lee describes the *kaross* as a "formidable one-piece combination garment-cum-carrying device that also does service as a sleeping blanket." It is used for carrying vegetables, nuts, water containers, and firewood, as well as babies.

Babies may have been the first things that hominid mothers carried, but they were not the only things it would have been advantageous to carry. In an article that looks at human evolution from the gathering point of view, Adrienne Zihlman and Nancy Tanner, anthropologists at the University of California at Santa Cruz, discuss how overheated, exposed, and vulnerable the formerly tree-dwelling apes would have felt on the open ground of the savannas. And especially apes that were pregnant or holding babies and so were required to move more slowly. Under these new conditions, hominids might use containers or carrying devices to acquire food as quickly as possible, then bring it to the safety and shade of trees and river banks to eat, a very different mode of consumption from that of other primates, including chimpanzees.

Chimpanzees eat most of their food where they find it. They carry food only minimally, to get it away from a harassing individual or to dangle a red colobus monkey in front of a hungry female. They eat when they are hungry, and they ignore food when they are replete. When they find a tree laden with fruit or nuts, they do not store any of the food for later. Such a seemingly simple thing as gathering food for future consumption offered hominids many advantages: the safety and comfort of being able to eat food under a tree *and* small excesses of food. If food is gathered to be eaten at some future point, it's hard to know exactly how much food should be collected, hard to know exactly how hungry one is.

We know what male chimpanzees do with an excess of food: they trade it for sex or influence. And given our understanding of what it means to be a male and the different reproductive strategies of males and females, that's just what we would expect males to do. Males are capable of fathering more offspring than females can mother, so selection in this sex has tended to reward promiscuity.

But what would a female do with an excess of food? She couldn't use it, as males do, to produce more children, since one child, every four years or so, is about as many children as a hominid female on the move can possibly handle. But she could use it to help raise her children. She could share it with her relatives and friends to create a more secure social context for her children to live in. She could use it to get more of what males have to offer: protection from predators and other males and occasional pieces of meat. By giving her excess food only to a male who protects her and provides her with meat, a female could use her excess food to shape male behavior. She could turn a male into a devoted mate. Males could also then afford to invest more of their time and energy in hunting, because even if they returned empty-handed, they could count on collected foods from females being available. By exchanging vegetable foods for meat and meat for vegetable foods, both males and females could ensure themselves and their children of a complete, balanced diet. By consuming increased amounts of meat and animal protein, they could better tolerate the chemicals in plants and so could eat more plant foods, deriving more benefits from their therapeutic properties. Food sharing is a social tool that the hominid female invented to allow her to find the resources necessary for her survival. It enabled males and females to develop another important social tool: the human division of labor.

But why wouldn't a hominid male, short on food, simply take any extra food that a female had? you might ask. The fossil evidence indicates that early hominid males (like contemporary human males) were larger than females. So why wouldn't they just usurp a female's food, as a male lion does?

The reason is that no species starts from scratch in its evolution, as Shirley Strum and others have noted. The behavior of primate males had already been shaped to the point where overt dominance tactics were not used by most males to get what they want. Monogamy is much more common in primates than in mammals in general, and even in nonmonogamous species like chimpanzees, males tend to remain in the vicinity of mothers and their offspring and to play some role in their protection and in the protection of the group as a whole. Chimpanzees do not form long-term pair bonds, but chimpanzee females tend to mate with males who groom them and who give them food.

This view of the human pair bond is a total about-face from the traditional view, the view most of us grew up with, in which men choose the women they will marry and support. But it is completely in line with the current understanding of how male and female reproductive strategies play themselves out in the real world. Males tend to be more promiscuous than females, but females, because they invest more in their eggs and fetuses and children, get to be choosy about who they mate with. And by being choosy, by carefully selecting their mates, they shape male behavior. If a female mates only with a male who protects her and brings her food, and that female raises more children than another female who mates with just any male (or just the strongest, most aggressive males), then pretty soon only those males that protect women and share food will get to mate. And pretty soon food sharing and defense will be part of the male makeup.

In early humans, pairing was aided and abetted by the ability of females to mate throughout their reproductive cycles and the loss of any external, visible signs of fertility. Chimp males can ignore females except at the stage of maximum receptiveness, which predictably accompanies ovulation, but in humans, as in all monogamous primates, ovulation is hidden or concealed. Sex in these monogamous species, which include the marmosets and tamarins of South America, is used for reproduction, of course, as well as to cement mutually profitable relationships between males and females, men and women. The human female's ability to mate throughout her cycle, and her strong sex drive, allows her to exchange sex, as well as food, for male commitment and paternal care.

There are, however, two important caveats to the ability of females to shape male behavior. One: females can't shape behavior in directions that males can't go. A ewe, for instance, can't expect food from a ram, because a ram has no way of carrying food. A small bird like a hummingbird can't demand much protection from its mate. Two: males are much more likely to stick around and help a mother to raise her young if they're pretty sure that they're the father of those young. Male behaviors like mate guarding, copulatory plugs, chastity belts, and—horrible to say—female circumcision enhance this sense of certainty.

We may never know for certain how important carrying devices and small excesses in food were to the story-arc of human evolution, but we do know the end of this story. We know that mothers did eventually tap into that great, previously untapped source of mother and child care, hominid males, and that males began to protect and support individual females and their children, thus giving rise to the nuclear family. We know that males began to concentrate on the risky, unpredictable business of hunting, and that females con-tinued to concentrate on gathering, a more predictable way of obtaining food. By providing their children with these different kinds of food, parents were able to feed their young long after weaning and to greatly reduce their risks of starvation. The hominid gut was able to shrink and the hominid brain to enlarge. Increased brainpower enabled us to develop increasingly sophisti-cated food-getting and food-processing techniques *and* to develop the intellectual and emotional capabilities that make long-term relationships—between parent and child, husband and wife, rela-tives and friends—possible. Diversification in feeding also allowed women to be less mobile, thus decreasing their risk of accidents and exposure to predators and increasing the amount of time they could spend caring for their young. With reduced mobility, women could shorten the interval between successive births and increase the number of children they could successfully raise.

It wasn't a quick transition, from ape to human, and it wasn't direct. Many different hominid lines—lines that ate the wrong foods and chose mates for the wrong reasons—evolved and went extinct. Two million years after the first members of the human line began walking upright across the African savannas, the brain began its great expansion. Two million years after that (about 200,000 years ago), Homo sapiens emerged. When it was all over, men and women needed each other and were well endowed with the physiology of love and attachment; children did best with the attentions and resources of two parents.

But for whatever reason, probably because the most devoted husbands don't always make the best hunters, men didn't receive a total makeover in this process. Their promiscuous tendencies often assert themselves. Women often choose success over devotion (I've heard this called the "I'd rather be miserable in a Rolls-Royce than happy in a Chevrolet" syndrome), and men are often able to have their cake and eat it too. In certain environments, marrying one woman and raising one family is a man's best reproductive bet, and in certain countries, laws and customs support this tactic. But in more forgiving environments, environments with more resources to be accumulated, men often take more than one wife. Marriage is a universal human custom, but eighty percent of human societies are classified as polygamous, a figure that doesn't even include societies like the United States, where polygamy is rare and officially illegal but serial monogamy is common.

So women often choose men with power (or resources) over men who dote on them; men often divide their time and resources between wives and families. That much is obvious and incontrovertible. But a corollary that many find harder to accept is that men in many societies restrict the amount of resources that their wives can use—even food resources. Food taboos are common throughout the world, in both hunter-gatherer and agricultural societies, and the most common taboos involve what and when women are allowed to eat. For instance, in many parts of southern Asia and Africa, women are prohibited from eating chickens and eggs—the same chickens and eggs that it is their job to raise. On the island of Alor in Indonesia, all flesh food or meat is viewed as the property of men. "Being women, eat crumbs" is a saying among the Chuckee of the northern extreme of Siberia, where women eat only after their husbands have eaten and have taken the choicest parts of the food. The sharing practices of the Australian aborigines—where the order of preference in food distribution is old men, hunting men, children, dogs, and women—blatantly prevent women from eating animal fat most of their lives. Even in households in modern-day Ireland and England, the choicest cut of meat is often reserved for the man of the house. When Virginia Woolf commented that "one cannot think well, love well, sleep well, if one has not dined well," she was referring to the great differences between the quality of food served in the men's and women's dining halls at Oxford University in the first part of this century.

To some extent these restrictions may be a reflection of the different taste preferences of men and women, since women, in the United States at least, are twice as likely to be vegetarians as men of the same age group. But in many countries, taboos and restrictions are known to deprive pregnant and nursing women of necessary protein and calories and are of serious medical and health concern. In South Asia, these food restrictions are associated with gender differences in mortality, morbidity, and malnutrition. In Asia, Africa, and Latin America, iron deficiency anemia affects two to three times more women than men.

Now some have argued that, because these restrictions exist, they must make a kind of nutritional sense. "Most men love, like or at least appreciate their mothers, wives, and daughters," observes Frederick Simoons, the author of *Eat Not This Flesh*, a comprehensive history of food avoidances and taboos. So despite evidence of links between gender differences in diet and health, Simoons contends that the chicken and egg avoidance common in Africa and Asia must have arisen out of concerns about the dangers of eating eggs. Others suggest that these taboos serve a beneficial function because they prevent women from consuming too much protein during pregnancy, overconsumption that could well pose a danger to fetal and maternal health if indeed it was an issue. But only in the Arctic and in certain industrialized nations where women choose to go on high-protein diets is the consumption of excessive amounts of protein during pregnancy a potential problem. And in the Arctic, pregnant women avoid this by eating a lot of seaweed, as well as the stomach contents of caribou and other herbivores.

Some have taken an entirely different tack, reasoning that because these food restrictions exist, human diets cannot be rational, in any meaningful sense of the word, that myth and symbol must be just as important in determining how people eat as nutrients and calories. It would not be rational or sensible for men to deprive their wives, and sometimes their children, of needed food, the argument goes. Since they do in many societies, human diets must not make sense. The fact that the egg symbolizes new life and is closely associated with sex and reproduction, for instance, must be just as important as its protein value.

But just because food restrictions don't make nutritional sense for women, they can still make sense in terms of the reproductive goals of males. It would be crazy and irrational, it is true, for men to deprive their wives and children of so much food that they sickened and died, but it might make sense (from the male's point of view) if they limited their wives' resources in order to have more of these resources to expend elsewhere—on other wives and other children. Food taboos are very old, but our understanding that they stem from fundamental differences between males and females is very new. It will take time for that understanding to sink in, time for us to discover what this means for us as we live out our long lives, loving, marrying, and bearing children with the same lopsided goals and tendencies, time for women to discover how best to safeguard their interests and those of their children.

Magdalena Hurtado, a Venezuelan-born anthropologist, learned about the food restrictions to which many women in the world are subject when she was foraging with a group of Ache in the tropical forests of Paraguay in the early 1980s. And she learned about them in a very personal way: when she became subject to the same restrictions as the rest of the Ache women.

"My memories of foraging trips are imbued with hunger," Hurtado writes in an account of the nine months she spent in eastern Paraguay with her husband, Kim Hill, and two other anthropologists. "I always felt as if I didn't have enough food; and I could never tell when I would get it. Between bouts of data collection I would oftentimes think about the two pieces of candy and a handful of nuts I had hidden in my knapsack for an afterdark meal."

Food became Hurtado's obsession when she was living with the Ache. This had something to do with her initial reluctance to eat monkeys, peccaries, armadillos, palm larvae, and the other foods of the forest (a reluctance she soon got over), but it also had to do with being a woman in a society where there are rules as to what, when, and where women are allowed to eat. Hurtado had come to Paraguay in 1981 with the idea that life in hunter-gatherer societies was egalitarian and that food was divided equally among all the members of a band. But belly first, she was beginning to question this view. Even though plant and animal resources are abundant year round in the forests inhabited by the Ache (forests that are very different in this respect from the forests in the Philippines or in Zaire, where the Agta and the Mbuti live), Ache women spent a great deal of their time thinking about food.

"Ache women were far more obsessed with food, especially meat, than the men seemed to be," Hurtado writes. Her husband was never as food-stressed as she was; nor was Kristen Hawkes, the other female anthropologist who was part of this study, and unmarried. As with the few unmarried Ache women, food was shared with her freely and plentifully.

Meat, Hurtado discovered one night when her husband was late in returning to camp, was not given directly to married women, but was divided among the men, who then gave it to their wives and whomever else they chose to share it with. Up to that point, Hurtado had been happy about the benefits of having a spouse in the field, but that night, while waiting for her husband, she looked on hungrily while everyone else, including Hawkes, enjoyed the meat that had been cooked.

Another time, Hurtado learned that rules also apply to food that she herself collected. One afternoon, after gathering the sweet fruits of the kurilla tree, she was so hungry when she got back to camp that she started to eat immediately. But "an Ache woman yelled across camp, with others joining in, that I stop eating the fruit. They said I had to save it for Kim and could only eat more after he had eaten. While ceasing to eat," Hurtado observes dryly, "I worked to convince myself of the merits of cultural relativism."

Eventually, Hurtado learned to manage her hunger by packing away any leftover meat that she and her husband had at night to eat clandestinely the next day, timing her snacks just right so that she would get to the meat before the maggots did. And she wasn't the only one to hoard food like this, she discovered by chance. Other Ache women also hid pieces of meat in their baskets and consumed them throughout the day. They, like Hurtado, kept their stashes well hidden from the prying eyes of both women and men.

You might think that women would band together to protect each other against these restrictions. But remember, women are in competition with each other, biologically speaking; they are competing with one another for resources. That competition can be keen in a promiscuous society like that of the Ache.

One of the questions that Hurtado and her colleagues were looking at during the time they spent with the Ache was how these hunter-gatherers share their different resources. How do they divide up and distribute the meat, honey, and vegetables they hunt and gather? They found a pattern to this sharing that is very different from the egalitarianism that hunter-gatherers were once thought to represent. Small collected resources, fruit, insect grubs, and the starchy pith of the palm (the abundant, high-carbohydrate food that enables the Ache to live independent of agricultural groups) are very rarely shared outside the family group. These foods are eaten by the immediate family— the spouse and the children—of the person who acquired them, most often a woman. But the meat of large animals and, to a lesser extent, honey are shared widely. So widely, in fact, that Ache women and children are no more likely to eat meat from animals killed by their husbands and fathers than they would be on the basis of chance alone. So widely, that most of the food acquired by Ache men goes to the wives and children of others.

The usual explanation for this pattern of sharing, found in a number of contemporary hunter-gatherer societies, is that the act of procuring meat is risky and unpredictable. Skill and luck are involved, and men are able to provide meat for their families on a regular basis only by sharing meat within a network of other hunters. This explanation is supported by the fact that the resources most likely to be shared *are* those that are the most unpredictable. It seems to be true, as far as it goes. But it doesn't go far enough. It doesn't explain, for example, why, if providing for their immediate families is their goal, Ache hunters usually pass up the opportunity to pursue small and easily caught game like armadillos in favor of the far riskier pursuit of large animals like peccaries. It doesn't explain why meat is shared less frequently outside the family in monogamous societies like the Hiwi, hunter-gatherers that live on the *llanos* or savannas of western Venezuela, than it is in promiscuous societies. Finding food is far less predictable on these savannas, which are flooded by heavy rains every year, than it is for the Ache. Yet most of the food that Hiwi men acquire goes to their own wives and children.

No, a full and complete explanation of the sharing practices of the Ache, and of hunter-gatherers around the world, must include the fact that meat and vegetable foods are shared differently because they are acquired by men and women, and men and women use their resources differently. Women use food to feed their families, and men use food to feed their families and, when they have the opportunity, to increase their mating possibilities. Hiwi men, who have a hard time finding food enough to survive and whose infants and children have a high rate of mortality, have few such opportunities. But in the forests of Paraguay, where both prey and palm fiber are available on a year-round basis, Ache men have plenty. This doesn't mean that all Ache men are hopeless polygamists, or that no Hiwi men divorce and remarry. Individual men differ in their ability to form attachments as in any other trait, and expression of this trait is influenced not only by the environ-ment but by any number of social and individual factors, including traditions, upbringing, character, talents, health, personality, and the affections of a particular woman.

But in a society where men have plenty of mating opportunities and where restrictions are routinely placed on women and their food consumption, Magdalena Hurtado, for one, was learning the importance of having her own stash of food (the equivalent, there in the Paraguayan forest, of Virginia Woolf's "room of her own") and of choosing her mate well. Her own husband gave her large portions of his food—though not the candy in his daypack.

S ometimes I wonder what I am telling my daughters when I stand before the stove preparing dinner for our family. What am I telling them about what it means to be a woman in America in the twentieth century and what it takes to juggle the roles of worker, mother, and wife—the same roles that women have been juggling since humans first evolved? What am I telling them about the lives that they might lead?

One thing, I hope, is the importance of taking care of the people you love and tending to their most basic needs. "No nation can be free when half the population is enslaved in the kitchen," Lenin is widely quoted as having said, but cooking and feeding others are acts of love (or they can be) and perhaps its first expression. Food also brings people together, and we are united as a family in part by foods prepared and partaken of in our kitchen—by the frequent risottos that we take turns stirring on the stove, by soups that change with the seasons and moods, by the perfume of sautéed chanterelles on toast. "O Soul come back to feed on foods you love" is the refrain of a Chinese poem written during the third or second century B.C. By filling my daughters with memories of foods tasted and savored and enjoyed in the company of our family, perhaps I am giving them a reason to keep coming home. Perhaps I am laying out a sensory path to our door.

But certainly, as we stand cooking and talking about our days, I am also telling them about the importance of choosing one's mate well. For the atmosphere in our kitchen, the taste and smell of the foods we eat there, is as much affected by my relationship with my husband, and by my husband's relationship with his daughters, as by the ingredients I buy in the supermarket. *Dry bread is better with love than a fat capon with fear*, the proverb says. *Unquiet meals make ill digestion*, Shakespeare observes in *The Comedy of Errors*. There are few societies in which a woman can live as fully as in the United States today, few societies that have more opportunities for women or where women's rights are more protected and guaranteed. Yet no amount of opportunities or laws can protect one from the consequences of one's choice of mate. A marriage can be a place where work is encouraged and accomplished, spirits are soothed and raised, lives are created and nurtured, and the essential differences between men and women are appreciated and talked about. A place where "your mind is your own and your heart is another's and therefore in the right place," to quote the very quotable M.F.K. Fisher once again. Or it can be a place where every instinct, idea, and emotion is thwarted and discouraged. It can be a heaven (at least some of the time) or a hell, a refuge, or a prison.

I was reminded of this not too long ago, when I sat next to Frank McCourt's wife at a dinner and I asked her why her husband had waited until he was in his sixties to write *Angela's Ashes*, his powerful memoir of growing up in Ireland. I half-expected Ellen McCourt to tell me that she had been the one to encourage her husband, since he had started the book shortly after they were married. But it wasn't that. It was more basic than that, more nitty-gritty, meat-and-potatoes than that. All Frank's previous relationships, his very open and engaging wife told me, had been so stormy and so full of conflict and drink, that he hadn't been able to write. Theirs was the first where he had some peace of mind.

So while cooking dinner, I am also telling my daughters about the importance of peace of mind, about creating a place—with good food too, that the soul wants to come back to.

Fourteen

EVERYDAY ANGELS

Magdalena Hurtado is not the only anthropologist to have learned as much about traditions of food sharing from her own personal experiences in the field as from the data she collected. In *Eating Christmas in the Kalahari*, Richard Lee's well-known account of the last Christmas he spent with the !Kung in 1964, he tells how he received his comeuppance for failing to share with the people he had made his subjects.

Unlike Hurtado and her colleagues, Lee and his wife, Nancy Howell, did not forage and hunt with the band of !Kung with whom they were living in the Dobe region of northwestern Botswana. Rather, they ate from a supply of canned foods they had brought with them. And they did not share any of this food with the !Kung because they didn't want to interfere with their food-gathering activities and dietary habits, the very activities they had come to study. But this practice couldn't help but cause a certain amount of resentment on the part of the !Kung, who rarely had a day's supply of food on hand. And the situation wasn't helped by the fact that Lee used small quantities of tobacco—the only tobacco available to the !Kung—to persuade these hunter-gatherers to cooperate with his collection of data. (The !Kung are no more immune to the pleasures and addictive quality of tobacco than people in industrialized countries.) "My approach, while paying off in terms of data," says Lee, "left me open to frequent accusations of stinginess and hard-heartedness. By their lights, I was a miser."

As the time for Lee's departure from the Kalahari approached, he thought to smooth over these resentments by purchasing "the largest, meatiest ox that money could buy" and giving it to the 150 !Kung in the vicinity for Christmas. He kept his eyes open for an animal of sufficient grossness and finally found one weighing about twelve hundred pounds on the hoof. Lee was certain that it would provide at least four pounds of meat for every man, woman, and child. Enough meat, he thought, to thoroughly satisfy everyone at the feast. Enough meat so that the trance dance, which follows every !Kung feast, would be a great success.

No sooner had Lee purchased the ox, though, than he began to receive visits from various !Kung complaining that the ox with which he wished to fête them was far too thin.

"Do you expect us to eat that bag of bones?" one woman asked.

"It is big, yes," acknowledged a hunter, "and no doubt its giant bones are good for soup, but fat is what we really crave, and so we will eat Christmas this year with a heavy heart."

The complaints were so numerous and so incessant that by the end of December, all the pleasure had pretty much gone from the gift that Lee had planned. He was even tempted to leave camp on the day of the slaughter so he would not to have to see the bickering that would take place when and if there was not enough

meat to go around. But the anthropologist in him made him stick
around and see it out. He and all the !Kung watched closely as the
man Lee had asked to perform the killing held his knife to the ox.
How quickly would it hit bone? How much fat was there really on
this enormous beast? The man made his first incision, then his second and third. The knife did not hit bone until the fourth or fifth
cut. The fat on the animal must have been two inches thick.

"Hey, that ox is loaded with fat," Lee burst out.

"Fat, you call that fat?" the butcher replied. "This animal is
thin, sick, dead!"

Then he and all the other !Kung started laughing. They rolled
on the ground paralyzed with laughter while Lee, whose Christmas had been nearly spoiled, tried to understand the joke. He was
still trying several days later, when he asked the same man why he
had told him that the ox was worthless. "It is our way," he told
Lee, adding that the !Kung are always dismissive about the qualities of any animal that is killed. That way a successful hunter will
not become arrogant and will not think of the rest of the !Kung as
his servants or inferiors.

"This way we cool his heart and make him gentle," another
hunter said.

"Why didn't you tell me this before?" Lee queried his many
informants.

"Because you never asked," one hunter answered, echoing a
refrain that Lee knew had haunted every field ethnographer.

Lee began to see the point of the lesson that the !Kung had
tried to teach him. Holding all the cards, as he did, all the tobacco
and all the medicine for miles around, "I was a perfect target for
the charge of arrogance and for the Bushmen tactic of enforcing
humility," Lee realized. The lesson hurt him greatly, for the black

ox was to be a totally generous, unstinting act. But, "as I read it," he adds, "their message was this: There are no totally generous acts. All 'acts' have an element of calculation. One black ox slaughtered at Christmas does not wipe out a year of careful manipulation of gifts given to serve your own ends. After all, to kill an animal and share the meat with people is really no more than Bushmen do for each other every day and with far less fanfare."

Humans, as I've said, almost always live in social groups in which some food is shared outside of the family. But in !Kung society, as in many groups of contemporary and recent hunter-gatherers, food is shared extensively among all those living together. When a giraffe is killed by a !Kung hunter, it is distributed among the group "until meat spread across the werf [camp] like a ripple across water," a visitor to the Kalahari in the 1950s noted.

Vilhjalmur Stefansson was the first outsider to report on the phenomenon of widespread food sharing among the Copper Eskimos of the Arctic after stopping in a village on the Dolphin and Union Strait in the spring of 1910 and noticing that "the little adopted daughter of the house, a girl of seven or eight, had not begun to eat with the rest of us, for it was her task to take a small wooden platter and carry the four pieces of boiled meat to the four families who had none of their own to cook."

Among the !Kung and the Copper Eskimos, food sharing is such an important part of being a member of a group that a refusal to share often precipitates the breakup of the group. Food sharing is so important and so extensive among these people that it seemed to anthropologists who came to live among them that food was a thing held equally by all, *that food was a common good.* "Sharing is such a pervasive feature of !Kung life that there is not much caloric advantage in being married to, or being the child of, a good hunter (or, for a man, being a good hunter rather than a poor or lazy one)," wrote Nancy Howell as recently as 1986, "since what matters is simply that there should be a good hunter in the group."

These anthropologists reflected on the role that exchanges of food play in fostering group adhesion and group survival, and they reflected that communism seemed to be the natural state of humankind. When Lee published labor statistics showing that the !Kung work only two to four days a week, with women working about a day less than men, they were quick to replace old views of the lives of hunter-gatherers as "nasty, brutish, and short" (in the words of Thomas Hobbes) with a new, glowing view of people living in leisure and complete harmony with nature and each other. "The original affluent society," as some began to call the !Kung.

But unfortunately, many of these conclusions were based on wishful thinking rather than close observation and on the erroneous though seemingly logical assumption that food sharing, or any other behavior, can be selected for by natural selection if it confers benefits on a group. Group selection, this is called, and it was a very popular theory until the 1970s, when Robert Trivers and other biologists showed that there is no way that natural selection can act on groups and group behaviors. It can only act on individuals and individual behaviors. So only if food sharing (or any other behavior) confers benefits to individuals will it be found in groups of individuals. Unraveling those benefits, as anthropologists like Hurtado, Hill, Kaplan, and Hawkes have been attempting to do with the Ache and the Hadza and other groups of contemporary hunter-gatherers, is leading to a richer and more complex, though somewhat less rosy, view of the lives of hunter-gatherers.

But even without the biological revolution of the 1970s, just how long could anthropologists, in their enthusiasm for the apparent egalitarianism of hunter-gatherer societies, ignore the many little inequalities that were always there for them to see: the fact that males often exert control over when and what females eat; or that the foods acquired by men and women are shared very differently; or that different hunters and their families get very different cuts of the meat that is being divided (and therefore very different contributions of proteins and fats); or that food is, in fact, not always shared. During the summer, when Copper Eskimos cooked outside their tents to avoid the smoke of wood-burning fires, everyone saw what was being prepared, and people circulated until all the food was eaten. During the winter, though, when food was cooked indoors over smokeless oil flames, women could and frequently did hide part of their food.

How long could anthropologists ignore the fact that bitter squabbles often accompany the division of a carcass and that hunters are, in fact, very concerned with the ownership of those things that they were fond of describing as common goods? John Marshall's short film *The Meat Fight* documents the lengthy argument that takes place when a young !Kung hunter loses track of an antelope he has shot and another group of hunters find and lay claim to the carcass, making clear that only a slow and careful determination of ownership and distribution can dispel the rising anger. Among the Copper Eskimos, distribution of those so-called common goods could be quite dangerous, as Arctic explorer Knud Rasmussen revealed in his 1932 description of the division of a bearded seal. The interests of the hunters, he said, "lie in cutting the parts they are to have as big as possible, and far into the next man's share. Consequently, the flensing proceeds almost like a free fight, like a knife duel, in which a man can cut right into the fingers of one at his side."

No, societies like the !Kung and the Copper Eskimos are not "communism in living," as they have been called. Nor are we fallen angels cast out by our capitalistic tendencies from a garden of shared delights. Sexual inequality is part of every human's life (the inequality that stems from the size of the egg and the sperm), and so is physical inequality, the fact that everyone comes into this world with different skills, physical attributes, personalities, and so on.

Even in societies where food is widely distributed, food is a private good, and individuals who own it accrue private advantages by sharing it. Some of these advantages are almost intangible—deference in decisions about travel, for instance, or an increase in the amount of surveillance that people are willing to provide for the children of a successful hunter, or the security that comes from having many people in your debt and willing to share their food with you. Others—such as mating opportunities or physical objects—are very tangible. When a !Kung has killed an eland, he takes shrewd note, as a hunter once told Lorna Marshall, of certain objects he might like to have and gives the owners of those objects particularly generous gifts of eland fat.

"The man who stands in the esteem of others," noted English anthropologist A. R. Radcliffe-Brown, of the Andamanese of the Andaman Islands in the Bay of Bengal, "is the skillful hunter who is generous in distributing to others the food he obtains."

When a !Kung hunter is particularly successful, "the people connected with him ate a great deal of meat and his popularity grew," Elizabeth Marshall Thomas once observed.

"You gotta think what you believe," the basketball player said, and until anthropologists stopped believing that hunter-gatherer societies were perfect utopias, they could not begin to think about what was really going on in these societies when food was being shared. They could not see the elaborate webs of reciprocation or the quests for prestige, status, and reproductive advantage. It was not as pretty a picture as the one in which a child brings food to all her neighbors, with no expectation of a return of favors. But it had the advantage of being more in line with the truth, and with truth, one hopes, comes better understanding.

Nor could anthropologists even begin to understand what might be going on in those societies of hunter-gatherers that do not share food. While the idea that hunting-gathering societies were characterized by the kind of extensive sharing seen among the !Kung was at the height of its popularity, to even suggest that there were hunter-gatherers who did not practice this same kind of food sharing was an anthropological heresy. But there are, and a lot can be learned from these exceptions.

Among the Hiwi of Venezuela, for example, a man and his wife forage and hunt together for all the small game and collected resources that their family eats. Among tribes of Tolowa, Tutuni, and the Yurok, American Indians who hunted and gathered on the northwest coast of California until the late nineteenth century, each family collected most of the food they needed. Except for large sea mammals—whales and sea lions—captured of necessity by groups of hunters, these Indians shared very little food. Wealthy men sometimes gave feasts to which all the members of the tribe would be invited, and one could look at these feasts as a redistribution or sharing of food. But they usually had a clear, ulterior motive. The men who gave them wished to build a large canoe or a new house, or they had their eye on some special object—a red-headed woodpecker scalp, for example, or a necklace of shell beads. So these feasts were accompanied by expectations of an immediate repayment either in labor or in prestige goods.

The Tolowa, Tutuni, and Yurok were hunter-gatherers like the !Kung. The foods they ate were the wild foods they found in their territories. But how differently they lived their lives and treated their neighbors. How differently they shared their resources, and how different, too, were the environments in which they lived.

In the resource-poor environment of the Kalahari, where !Kung hunters might have to follow a wounded giraffe for five days after it had been shot with a poison arrow—five days of walking some thirty miles a day before the wounded giraffe was sufficiently exhausted that the hunters could finish it off—sharing is necessary to protect one against the desert's extreme unpredictability. Sharing may give a successful individual hunter added benefits, reproductive and otherwise, but it is also necessary to put food on the table every day, to reduce the daily variance in a family's food intake. When the successes of a group of !Kung were actually tracked over a period of a month, only one hunter was found to have been successful in taking any large animal.

The reason why the !Kung work so little, as anthropologists who followed Richard Lee to the Kalahari learned, is not because life is easy in the desert, nor because the !Kung, with their detailed knowledge of their surroundings, can, in just a few days, reap all that they need to survive. Nor is it because the !Kung do not want to deplete their resources, as has also been suggested. The !Kung work so little because life there is so hard that they must be careful about expending more energy than they can take in. (The !Kung are small in stature, as already noted, and it has been argued— quite vigorously by some—that small size must have some advantage in the Kalahari Desert and must have been selected for over time. More recently, however, anthropologists have been attributing the small size of the !Kung, at least in part, to chronic caloric insufficiency. The abundance of mongongo nuts, the staple food that provides one third to one half of their daily calories, means that the !Kung are never on the brink of starvation, but they are food stressed and lose weight for part of every year. They also have low fertility rates and high rates of infant mortality.)

Security, there in the Kalahari, as in many marginal environments, is having as many people as possible who are obliged to share what they have with you. Security is a web of reciprocal obligations woven of kinship, friendship, the exchange of arrows, and so on. !Kung tradition has it that the hunter who owns an animal (and therefore is in charge of its distribution) is the owner of the arrow with which the animal was shot. !Kung men frequently exchange arrows, thus forming a kind of Kalahari lottery in which a man may "win" an animal even though he may be far away when the animal is killed.

In the Arctic, where hungry bears dream of whales, men have even more elaborate customs for establishing meat-sharing obligations, including adopting children (even from two perfectly fit and loving parents), wife-swapping, and seal-sharing partners. Ringed seals, the staple food of coastal people in the Arctic, are shared according to *piqatigiit*, an orderly system in which men agree to become "stomach meat companions" or "neck companions" or "lower spine companions" and agree to give each other those parts whenever they kill a seal. Each man, in the highly unpredictable environment of the north, can have as many as twenty sharing partners. (Ringed seals are the staple food, the bread and butter of the northern diet, and are shared according to this harmonious system, but bearded seals, a much rarer find, represent a windfall, and their division proceeds, as Knud Rasmussen observed, much more violently and haphazardly.)

In the richly diverse environment of the northern coast of California, on the other hand, an abundance of acorns, shellfish, surf fish, land mammals, salmon, berries, and edible plants like the bulbs of lilies (the bulbs that Mark Brest van Kempen ate during his month in the wild) meant that food was plentiful for the Tolowa, Tutuni, and Yurok almost all year round. Food sharing was not necessary to prevent variance in food intake. There, where food could be preserved, by drying and smoking, for the brief lean period that the Indians experienced during the few weeks before the salmon started to run, families could gather all the food they needed for their own consumption. Whatever sharing took place became a way of obtaining things other than protection against starvation. Polygamy, not surprisingly, was common in these tribes, and a man with several hardworking wives and daughters could store up large reserves of food with which he could acquire more wives and give feasts to pay for more labor and goods. Though men performed the heavy jobs in these tribes, women were the primary producers, and the amount of food a family could accumulate was directly dependent on the number of women in the household.

We tend to think of hunting and gathering societies as being mobile and of having to move through large territories in search of sufficient resources. But there on the coast of California, resources were so dense that Indians lived in semipermanent settlements.

We tend to think of hunting and gathering societies as having little social stratification, of every man having a say in the decisions that affect a group. But in the tribes of the Tolowa, Tutuni, and Yurok, "big men" could accumulate not only wives, children, food, and wealth, but status, prestige, and political power.

"A bird can roost on but one branch," says a Chinese proverb, "a mouse can drink no more than its fill from a river." But humans are different. Humans can bottle the river and occupy the whole tree. Think of it. Only because men and women share food on a daily basis can some individuals totally free themselves from the task of finding food and turn their attentions elsewhere. Other animals feed themselves, but because men and women adopted a survival strategy that entails regular food sharing and a division of labor, some no longer need to perform this basic task. By dint of hard work (or hardworking wives), talent, force of personality, luck, and/or the nature of the environment in which they live, some can get others to forage for them.

Food sharing and the division of labor, Sarah Hrdy writes in *The Woman That Never Evolved*, carries with it the potential for one person to benefit from another's work, altering both relations between the sexes and relations between men. There's nothing new about males controlling the reproductive output of as many females as they can. That's what males do throughout the animal kingdom. What's new about humans is that men can control both large numbers of women *and* the resources they need to reproduce. What's new about humans is that men (and sometimes women) can control the resources needed by other men and women to reproduce.

"Hunger tames us," George Eliot has one of her characters say in *Middlemarch*, and though she was speaking of emotional hunger, this is true, equally, of the physical variety. And true, probably, of every living thing, though only humans can take advantage of it. Because we share food, food can be used to control others. *He who has no bread has no authority*, a Turkish proverb says. Chimpanzees show the most rudimentary beginnings of using food as a form of power when they share meat with certain individuals in order to cement an alliance or to obtain a sexual favor.

But because chimpanzees find their own food most of the time, because chimpanzees are never dependent on each other for food, food usually stays food for chimpanzees. It rarely morphs into power.

"The radical nature of this [the human] cooperative dietary strategy remains underappreciated," says Katharine Milton. It provided dietary benefits that were responsible for the expansion of the human brain, and it has social, emotional, and political ramifications that we are still trying to figure out.

For instance, only because we share food does it make a difference if food is accepted or refused. No chimpanzee or lion would even notice if another chimpanzee or lion refused to eat, but anorexics, hunger strikers, and picky three-year-olds affect others greatly by not eating, by denying themselves food. Refusing to eat is an exercise of self-control. It is also an exercise of control over others and an equally powerful weapon, says anthropologist Pat Caplan, in the hands of Gandhi or a teenage girl. An angry husband often refuses the food that his wife has prepared for him. In India, some women seem to voluntarily deprive themselves of food. They acquire respect and show their devotion to their family, by the act of going hungry.

Only because we share food do people define themselves by the foods they eat and the foods they don't eat, do foods have the power to mark social, religious, ethnic, and class boundaries. "Speak not to me with a mouth that eats fish," Somali nomads used to taunt outsiders. "Is their mouth the same as ours?" Tibetans ask of other groups. Strabo, the classical Greek writer, identified people living south of Egypt as Chelonophagi (turtle eaters), Elephantophagi (elephant eaters), Spermophagi (seed eaters), and Acridophagi (locust eaters), in the same way that Germans are or were known as Krauts and the French as frogs. Groups are most often defined by the meat or flesh foods they eat, but in China, the people of a village in Shantung are referred to as sweet-potato eaters, an offensive term because sweet potatoes are deemed an inferior food and eating them is a mark of poverty. In India, rejection of beef-eating may have had its origins among the Brahmins as a stratagem in their extended struggle against the Buddhists. In the United States and Europe, vegetarianism has links with feminism and the questioning of male-dominated societies.

And only because we share food with each other can we share food with the members of other species, can we have pets and zoos and raise domesticated animals. I think of Ellen Dieren-feld as I write this and her puzzlement over why humans tend to overfeed everything, including animals in zoos. Dierenfeld was right: feeding others is a part of human nature, a part of us ever since women and men began providing each other with food. But there's an irony here that I must talk to her about the next time I see her. This human practice of sharing food is what makes zoos possible. For without it, how could we even conceive of a zoo? How could we think of feeding animals instead of eating them? But it is also what makes zoos necessary as human popula-tion growth, the result of this highly successful adaptation, threatens the habitats and existence of more and more animals around the world.

"Sharing food is an intimate act that should not be indulged lightly," observed M.F.K. Fisher. Fisher wrote eloquently and evocatively about the complex relationships between food, cook-ing, and love, but could she have known how profoundly she spoke? That human life, in all its astonishing complexity and vari-ety, sprang from this original, intimate act, repeated night after night, in house after house, hut after hut: women collecting and preparing food for their husbands and families, husbands bringing food back to their wives. Because humans share food, we can exchange it for other resources, and we can make our livings from resources other than food. We are even able to use each other as resources, as priests use believers, actors use audiences, and writers, readers.

No wonder humans are both so competitive and so cooperative, both so giving and so taking. We compete for resources, yet many of us rely on others to be the resources we need to survive. Or we rely on others to buy resources from us, exchanging them for money or other goods. No wonder humans are so social and have such sophisticated means of communication. When individuals gather and eat food by themselves, social interactions may influence their behavior. But when they collect food, which they later exchange with others, these relations become all-important.

Because we share food, we can recognize hunger in others and can give to others in need. We can respond to the news of a famine or a friend's illness with outpourings of food. We can set up soup kitchens and create food stamp programs. We can use food to express compassion, friendship, gratitude, and hospitality, as well as authority and status. "Is virtue a shared box of chocolates?" asks Matt Ridley in his book *The Origins of Virtue.* The answer is probably yes. Sharing a box of chocolates or a stew of carbohydrate-rich vegetables and protein-rich meat gave humans the ability to see beyond themselves, *and* it allowed them to acquire excesses of food, excesses that they could then use in many different ways.

Food was never a simple thing for the omnivorous primate that we are. But once we adopted this system of food sharing, it became a kind of Chinese puzzle, a passage to India on which some humans would be freed from the task of finding food and food would become a language with a complex grammar and a rich, inexhaustible vocabulary. A woman can show her affection for her family by preparing them scrambled eggs for breakfast. Gertrude Stein's cook Hélène could rebuke Matisse for dropping by unexpectedly at mealtimes by making him scrambled eggs instead of an omelet. She broke the same amount of eggs and used the same amount of butter, but she left out cuisine, a subtle gastronomic scolding that was understood by all.

Fifteen

CONVERSATIONS WITH BONES

Like many people, I suspect, I am drawn to the sight of a field of wheat or rye, to an orchard of well-pruned apple trees. I enjoy their tidy opulence, their reassuring sense that all is right with the world. Certainly, wherever one sees such sights—even when the wheat and apples belong to some unknown, unrelated person—one expects to eat well at the end of the day. Fields and orchards speak to us in the most basic language there is. *Here, there is no want,* they say. *No hunger. Here we are well fed. Here we don't have to think too much about food. Our songs, unlike the songs of the !Kung and the songs of the Inuit, need not be about the search for food. They can be about love and not about hunting, about love lost and not about finding bitter melons and squeezing them for their pulp,* as I once heard a !Kung sing.

"The image of a market overflowing or a garden overgrown may appease our innate fear of hunger which surely lingers in some dark, primordial cranny of the human psyche," as food writer

Molly O'Neill observed in one of her columns. I like to forage in the woods for certain foods and their wild, *unpurchasable* taste, for the sense that I am reaching into a vitamin grab bag with every bite. But like most everyone living in the twentieth century, I depend upon agricultural crops to fill my belly. And I like to see those crops in the ground. My husband and I drove our daughter out to St. Louis recently, and for me at least, the fields of corn and soybeans made the miles pass more quickly.

Such is the effect that agriculture—its fields and crops, its abundance and order—tends to have on us. One might worry about the use of pesticides in commercial agriculture today or the effect of monocultures on biodiversity, but who would ever think to question the value of agriculture itself: agriculture, that new way of procuring food first devised by humans some eleven thousand years ago.* It was, people say, our great leap forward, the advance that catapulted us out of the hand-to-mouth, day-to-day existence of hunter-gatherers like the !Kung and Ache and into the complex, cultured, literate existence of modern human beings. Soon after groups of humans began to domesticate plants and ani-

---

*Agriculture was new to humans eleven thousand years ago, but not necessarily new to the planet, since we are not the only animals to raise animals and grow plants.

Leaf-cutter ants of the rain forests of the New World strip leaves, not to consume themselves, but to feed to the fungi that they do eat. And in an interesting parallel to slash-and-burn agriculture, bell miners of the eucalyptus forests of Australia protect the insects that feed on those trees, psyllids, from other birds, eventually causing the trees to die. Other birds eat the entire insect, but bell miners eat only the insects' lerp, a sweet carbohydrate shell that they grow to cover themselves and later shed, leaving the insects to destroy the trees.

"Bell miners are true farmers," says a biologist who has studied the activities of these birds; "their territoriality turns the trees into producers of bird food via psyllids, but at the forest's expense."

mals, they began to write down their thoughts, and to build their cities, and to live the life that we associate with being human.

Clark Spencer Larsen, an anthropologist at the University of North Carolina who comes from a farming family in Nebraska, was raised on this benevolent view of agriculture, and he also remembers well the archaeology textbook that he had as an undergraduate in which he read that hunter-gatherers lead a savage's existence and that agriculture was the beginning of all good things. He fully expected to find evidence to support this view when he began looking at human fossils from burial mounds on Saint Catherines Island off the coast of modern-day Georgia, where local populations of Indians began growing corn around A.D. 1150. What he found instead, as he examined fossils from mounds dating back three thousand years, was a far less bullish picture. It is, though, the same picture that has emerged in study after study of the same transition from hunting and gathering to agriculture all over the world, from far-removed locations in Europe, the Americas, and the Near and Far East.

When Larsen compared human skeletal remains from nineteen preagricultural and fourteen agricultural mortuary sites, it became clear immediately that the bones of the early agriculturists are very different from the bones of the hunter-gatherers that preceded them. (Larsen first identified the sites as being either agricultural or preagricultural by the presence of ceramics and other objects, but he later verified that they belonged in one or the other category with new methods of bone chemistry analysis that detect the carbon traces left by a diet of corn.) The shinbones of the farmers are sometimes swollen and pitted, evidence that they were suffering from periostitis, an inflammatory response to a long-term infection that had eventually spread into the bones. The incidence of these kinds of infections, infections like those that cause syphilis or tuberculosis, rose by eleven percent when corn began to be cultivated. It went from four and a half percent in preagricultural times to fifteen percent in the period soon after the advent of agriculture.

Growth and stature were also profoundly affected by agriculture. The farmers were greatly stunted; the long bones of their arms and legs are shorter, sometimes by as much as three inches, than those of hunter-gatherers. It used to be thought that the average height of humans has increased over time. But there is now much evidence to suggest that just the opposite has happened and that many agricultural populations have been dwarfed until the present century, when improvements in the growing, storage, and shipping of food have improved nutrition for some, providing adequate amounts of protein so that some peoples could grow to their full height. Average height estimates for the Cro-Magnons of France, hunter-gatherers who lived 25,000 years ago, are five feet ten inches for males and five feet six inches for females, and those averages are seen now as representing the genetic capability of our species. During the Revolutionary War, American soldiers averaged just five feet five and a half, but American men now average five feet nine inches. Because of our diet, Americans are now near or at their genetic potential. (Stature can change even more quickly, I learned on a trip to Japan, when I was walking past one of Tokyo's famous cram schools. Outside, there were groups of teenagers, some of them five feet eight, nine, or ten. Yet many Japanese from the prewar generations are no more than four feet tall. I wondered how Japanese society was adapting to these new Jacks in the Beanstalk and what effect this adolescent growth spurt was having on architecture, clothing manufacture, and relations between the generations.)

The teeth of the early farmers on Saint Catherines Island, one of a string of islands in the embayment known as the Georgia Bight, are in particularly bad shape. They are much more likely to be occluded or offset than the teeth of the earlier hunter-gatherers—and are much more likely to have cavities, holes ranging from the size of pinpricks to a loss of the whole tooth structure. Before farming, only about ten percent of the Indians had cavities; after farming, sixty percent did. "Orthodontists and dentists would go out of business if they just had hunter-gatherers to deal with," Larsen observes. "These people didn't have cavities; they didn't have overbites; they didn't have crooked teeth. We see these things developing because of the change in the diet and the way food was prepared. Corn had to be ground and cooked to be digestible, and soft sweet foods stick to the teeth."

"For much of intellectual history," Larsen told me when I visited him at his offices at the University of North Carolina in Chapel Hill, "the popular and scholarly perception of agriculture has been that once it was acquired, Homo sapiens had it made. Life improved dramatically, from a state of incessant work to abundant leisure, from deprivation to plenty, and from sickness to health and increased longevity. The burden of constantly moving about the landscape in search of food was replaced by the security of settlement in permanent towns and cities, from which the many advantages of civilization could be enjoyed. This data from Georgia, though, and similar data from sites all around the world, is changing this perception." The only place, in fact, where a switch to agriculture has *not* been associated with significant health problems is a site in Africa's Nile Valley (Wadi Halfa), where farmers ate their grains fermented and the grains thus acted both as natural antibiotics and potent sources of essential B vitamins usually found only in meat.

It is not that hunter-gatherers do not have any health problems or suffer from infectious diseases, says Larsen, a boyish-looking, genial man in his late forties. Or that all of their diets are ideal. But as a group their diets were better balanced in broad composition than those of most agricultural or industrial populations. By consuming a wide variety of foods, they were better able to satisfy their nutritional requirements than individuals in specialized farming populations. It has often been observed that groups of contemporary hunter-gatherers are better nourished than their neighboring agriculturists.

Corn is behind many of the problems that Larsen sees in the early farmers because corn is low in protein or available protein and therefore a poor food on which to grow and with which to fight infections. Corn, in addition, blocks absorption of iron from other foods and can cause anemia. And it is a soft food that is high in sugar and causes tooth decay. In other parts of the world, rice, barley, and wheat have caused similar constellations of medical problems, so corn is not unusual in its ability to detract from health. Overreliance on any one food source is the real problem for an omnivore like Homo sapiens.

Larsen's data from the Georgia Bight is changing perceptions about agriculture, and it is also changing perceptions about gender issues in our distant past. Anthropologists used to think that gender was a largely inaccessible part of archaeology, but Larsen's very large samples of bones, representing a time period of three thousand years, have made gender, he says, "a highly visible part of the past." His studies have opened new windows on the relations between men and women long ago. In Larsen's samples, women in agricultural societies got more cavities and experienced more tooth loss than men, and their loss in stature was much greater. At the same time, they put on much more weight, a trend that Larsen has deduced from the greatly increased strength of their bones. In order to consume adequate amounts of protein from a diet of corn, a person would have to eat many pounds of this high-calorie, low-protein food each day, leading, of course, to weight gain and to bones that have to grow stronger in order to support the increased body mass. Agriculture is many things to many people, but archae-ology is showing that it may also have been the beginning of female concerns with dieting and weight.

"Agriculture had an impact on both sexes," Larsen points out, "but its impact on women was far greater." His findings indicate that women were eating more corn than men, and that they were eating less of foods that were high in protein. Larsen believes that the women farmers of Saint Catherines Island consumed more corn than men because they spent more of their time growing corn and preparing it for consumption and therefore had greater access to corn. But given the food taboos that exist in many societies and the widespread restrictions on protein intake by women, perhaps we should be a little suspicious of this explanation. It may be too simple. Agriculture, with its greater opportunities for creating and controlling food surpluses, could well have amplified inequalities in the diets of men and women that were already well established. Women may have been eating more corn than was good for them because men were not freely sharing with them the foods they had obtained by hunting and fishing.

Larsen also attributes the differences in cavities and tooth loss between the sexes to the different eating *styles* of men and women—and here he makes an interesting point. Tooth loss in women used to be blamed on pregnancy, and I remember my mother-in-law telling me that a woman loses one tooth and gains one shoe size for each of her children. But "that's an old wives' tale," according to this bioarchaeologist. "Teeth are stable entities and do not give up minerals during pregnancy or any other time." This old wives' tale, though, probably had its origins in the different diets and food habits of men and women. Men tend to crave salty or meat-containing foods; women, sweet foods. Women, it has also frequently been observed, tend to eat more small meals during the day, and men, fewer but larger meals, a contrasting meal plan that probably has to do with the fact that men hunt and women gather in most societies. Gathering, of course, lends itself to snacking, while hunters often have to wait a long time for their meals. And this female way of eating can itself have a negative effect on teeth. Bacteria are released every time we eat, so the more times we eat during the day, the more times our teeth are subject to the agents of decay.

An overreliance on corn is responsible for many of the health problems associated with the switch to agriculture on Saint Catherines Island, but corn is not the only culprit. Larsen also blames higher population densities and permanent settlements. As settlements on Saint Catherines grew, Larsen notes, "people were living right on top of their garbage and often contaminated their drinking water." As people live closer and closer together, the incidence of infectious and parasitic diseases also rises.

For there is a curious paradox in Larsen's picture of the life of these very early American farmers. Despite their high rates of infection and their rotting teeth, despite their poor nutrition and their loss of stature, their population was booming. The small, temporary shelters of hunter-gatherers who had wandered the Georgia coast in search of seasonal foods had become large permanent settlements, and the permanent settlements grew as agriculture made an ever-more-significant contribution to the diet. It used to be thought that if a diet didn't provide just the right combination of nutrients, a population would die off, but here in Georgia, and in other sites marking the transition to agriculture, health problems were not enough to keep populations from growing.

But why would a population switch from the diet of hunter-gatherers, a diet that provided it with a healthy variety of foods, to a diet much more likely to result in malnutrition and disease? Why would the people of Saint Catherines, living in such a rich, productive part of the world, trade in their digging sticks and arrows for plows and grinding stones?

I asked Larsen that question while we were eating lunch in a restaurant near his office in Chapel Hill. We had both ordered, ironically enough, open-faced corn tortillas with vegetables and beans, a New World concoction made possible with the advent of agriculture but made completely nutritious with a sprinkling of cheese on top—a topping not available to Indian agriculturists since they did not keep domesticated animals. I hadn't yet read Mark Cohen's influential and eye-opening book *The Food Crisis in Prehistory*, and I had no idea of what deep waters I was getting into, the controversy that surrounds any discussion of human population growth.

But before I let Larsen respond, I tried casting my own mental nets around for the answer. "Was it because agriculture gave agriculturists more security and less variation in their subsistence base?" I asked him. "Did farmers face less risk of hunger and starvation?"

Probably not, he told me, for there are many indications that hunter-gatherers in many different locations fare better than agriculturists during droughts, famines, and other environmental changes. Their greater mobility and their diverse diet enable them to minimize the effects of natural fluctuations in their food supply much more easily than sedentary food producers. During hard times, in fact, many marginal agriculturists have been observed to switch back to hunting and gathering. In the twentieth century, with more advanced means of growing, harvesting, shipping, and storing food, agriculture *can* protect against want during times of scarcity. But that security was a long while in the making. It would not have been part of the motivations of the first farmers.

Then what about the workload of the two lifestyles? "Was early agriculture an easier way of life?" I asked. "Would hunters and gatherers switch to it because it was less stressful and gave them more time to do other things?" Larsen has found that the transition to agriculture on Saint Catherines Island *was* associated with a marked decrease in osteoarthritis, a condition that manifests itself as either a breakdown or a buildup of bone and can have a number of causes, though predominantly the heavy, repetitive use of joints. Decreased osteoarthritis is the only medical benefit associated with this new way of life. But he doubts whether this benefit actually translated into greater leisure since time budget studies show that most peasant farmers and herders spend more hours at work than hunter-gatherers.

What was left then, if health, security, and leisure weren't incentives for agriculture? There must be some reason why humans abandoned hunting and gathering and abandoned it en masse—within a period of just a few thousand years. Eleven to twelve thousand years ago, all humans were hunters and gatherers and lived only on the wild foods they were able to find. By 9000 B.C. or so, most humans on the planet were farmers or pastoralists. Populations in the Americas hunted and gathered for several more thousand years, but only in a few geographically isolated areas like the Kalahari Desert and the forests of Paraguay have hunter-gatherers persisted into the present.

The answer, when it came, was so simple that it took me a long time to absorb it.

It was so simple that I thought, at first, it required further explanation, and in the notes from my conversation with Larsen, I can see that I took him around and around in circles, asking the same questions in different ways. I wasn't able to accept that he had already told me what I wanted to know.

The answer to this question, as to so many questions concerning the behavior of humans, and of all animals, has to do with babies. Hunting and gathering, as Larsen explained, is the most successful mode of human existence both in terms of the length of time that people relied upon it as a way of life and in terms of the health of its practitioners. If it wasn't such an effective dietary regimen, humans and their ancestors couldn't have used it to survive for so many thousands and thousands of years. Its only problem was that it was too successful. Humans throve on it. They multiplied and divided on it. And when their numbers became too great for an area to support, they migrated to new places all across the globe. There they continued to practice their division of labor, adapting it to new environments as they moved, so that in some places, like the Arctic, women no longer foraged but performed essential work making clothing and preparing food, while in others, like the forests of the Philippines, women hunted alongside men. Hunter-gatherers, we now know, are incredibly flexible in these kinds of behavioral details. They do whatever is necessary to keep themselves fed.

But by twelve thousand years ago or so, humans had occupied all those parts of the earth where they could support themselves by this method of finding food. There were no more unoccupied places for them to migrate to, and much pressure was put on existing food resources. Some of these resources, like the mastodon, they ate to the point of extinction. Others—insects, shellfish, the grains of cereal grasses—they exploited for the first time, expanding their resource base during these times of scarcity in a way predicted by optimal foraging theory. Given the current dependence of human populations on cereals and the fact that some human populations obtain ninety percent of their calories from wheat or rice, it is curious that no hunter-gatherers in Africa collect wild seeds. Nor do any African monkeys and apes. But not that curious. Gathering seeds takes time and requires lengthy winnowing and grinding procedures. The only hunter-gatherers to exploit seeds are Australian aborigines, who have eliminated some of these drawbacks by raiding the underground grain stores of harvester ants. By this unscrupulous technique (at least from an ant's point of view), aborigines managed to make this labor-intensive resource profitable—at least for a time. When Australia was colonized by the English and wheat flour became available, they soon abandoned this technique.

Hunter-gatherers in Europe only began eating great quantities of seeds when other resources grew scarce and when new technology— sickles, baskets, grinding stones, cooking and storage techniques— made the collection and processing of seeds profitable. Evidence from many different locations indicates that humans had considerable experience with intensive grain collecting (or maize collecting in the Americas) long before plants were cultivated or domesticated. Hunter-gatherers in the Fertile Crescent (where einkorn wheat was first domesticated around eleven thousand years ago) had even managed to settle down into permanent villages, sustained by the amount of seeds that they were able to collect from natural stands of wild wheat.

Then, when populations continued to grow and food supplies once again became a problem, some had success in feeding their families by encouraging these cereal grasses to grow through the controlled use of fire; by selecting for certain plants with

desirable mutations: grasses that produce seeds with a higher protein content, for instance, or grasses with stalks that don't shatter
as easily and that thereby facilitate harvesting. It used to be
thought that agriculture began at the end of the Pleistocene Era
because retreating ice sheets stimulated new mutations in plants
and provided places for them to grow, but the opinion now is that
mutations are always being produced. What changed at the end
of the Pleistocene was our willingness—our need—to take advantage of them.

"The thing about agriculture," the patient Larsen said as we
ate our tortillas and circled around and around this idea that agriculture is the result of population pressure, "is that it feeds more
mouths per unit area of land than hunting and gathering." Population densities of hunter-gatherers are on the order of one person
per square mile; densities of early farming populations were sixty
persons per square mile. An acre of wild, uncultivated land yields
an edible biomass of one tenth of a percent; an acre of cultivated
land, a biomass of ninety percent. Raising a goat or a cow for milk
instead of meat means that the goat or cow yields many more calories over the course of its lifetime.

The lightbulb was beginning to go on.

Simply put, during that time when the earth seemed to have
reached its carrying capacity for the big-brained, two-legged creature that had successfully populated all its lands, those people who
took the first small steps toward raising crops and domesticating
animals raised more children than those who persisted with the old
ways of finding food. These weren't conscious, planned steps.
It was just that those hunter-gatherers who happened to encourage the growth of plants or the domestication of animals raised
more children than those who did not. Many more children, as it

turns out. The reproductive rates of agriculturists—even the first agriculturists—can be one hundred times higher than those of the earlier hunter-gatherers. These children weren't as healthy as those raised on the wild foods of the past (foods that provided a much broader array of nutrients *and* beneficial plant secondary compounds), and they weren't as tall, but they survived and reproduced and learned to raise food the way their parents did. Because there were so many of them, agriculture very quickly took the place of hunting and gathering in all but the most marginal and/or remote areas of the world. Only in those places would hunting and gathering persist as the better strategy.

Throughout human existence, evidence now suggests, food production and population density have been linked together in a positive feedback cycle in which gradual increases in population have impelled people to find more food and more food has led to increases in population. Once people settled down and stayed in one place—either by producing food or by living in a part of the world where resources were so rich and diverse that it was possible to become sedentary as a hunter-gatherer (the northern coast of California, for example, or the coastal regions of Japan), they could shorten their birth spacing and produce still more people, requiring still more food. "In most animals, population is constrained by the food supply," Larsen explained. "Most animals cannot use technology to overcome this constraint, and so their population tends to remain the same over the long term. Humans, on the other hand, respond to population pressure with new ways of expanding their food supply."

As Larsen said this, I thought back to the simple technology of chimpanzees that allows them to open nuts and to probe for ants and termites, and to a comment that Japanese primatologist

Toshisada Nishida made in reference to the debate over whether animals other than humans can have culture, defined by anthropologists as the social transmission of acquired behavior. "The significance of cultural behavior," this thoughtful scientist said, "lies primarily in the fact that it opens a new ecological niche not exploited by other animals." Many of us think of culture as a by-product of civilization, "the intellectual side of civilization," as the Oxford English Dictionary defines it. But if culture, like technology, opens up new ecological niches, *if it is an expression of intelligence rather than a product of civilization*, it would have been a part of the very earliest human life. And though most people would prefer to keep culture as their own exclusive club, why wouldn't culture, like technology, be present in chimpanzees, and certain other highly intelligent animals, in some rudimentary form?

Culture and technology don't just enable populations to grow, of course. They enable them to grow in new directions. With increased agricultural productivity, more and more people were freed from the task of producing food and freed to specialize in other endeavors, in the opening up of new *cultural* niches and the accumulation of other kinds of resources. Some became warriors, for the same population pressures that led to agriculture also led to an increasing emphasis on warfare, and some began to control the food produced by others. Everywhere, the rise of agriculture is associated with the appearance of chiefs and with increasing signs of privilege and status, including vast differences in diet according to class. The Sumerians of Mesopotamia left detailed pictorial records of the food rations accorded to various classes, showing that the lower strata of Sumerian society relied heavily on barley, while the higher classes had a much more rich and varied diet. The bones of individuals buried at Chalcatzingo, Mexico, during the years 1150 to 550 B.C. have in them different levels of strontium, a chemical that is indicative of the amount of plant food in the diet. Those individuals with low levels of strontium in their bones, those, in other words, who had eaten more meat over the course of their lifetimes, were found buried alongside elaborate grave goods, an indication of their wealth and social standing. Those with high levels were found in common graves. So to a bioarchaeologist, the saying "you are what you eat" is literally true. Many foods have chemical signatures that come to be incorporated into the skeletal system and can be read thousands of years later to reveal the nature, and the inequities, of the foods that we eat.

If hunter-gatherer societies were truly egalitarian, as they used to be portrayed, it would be difficult to account for this sudden change to class-based societies. Blame for this change has often been put on the growth of technology or the surpluses produced by agriculture. But knowing that the quest for status and influence and reproductive advantage has always existed, no finger-pointing is necessary. Agriculture magnified the extent of social and political stratification that was possible, but stratification did not begin with agriculture. In the sedentary hunter-gatherer societies of the Indians of the Pacific Northwest or the Jomon of Japan, thought to be the first people to make pots out of clay, no crops were cultivated, but archaeologists have found many objects reflective of personal status and privilege.

There is much evidence to support this population-driven view of human history to which Clark Larsen was opening my eyes that day over lunch, though some of it is still controversial. It includes the sudden mass extinctions of large mammals in the Americas and in Australia–New Guinea shortly after the time when humans are thought to have first arrived—14,000 years ago in the Americas; 40,000 to 30,000 years ago in Australia. Fifteen thousand years ago, the American West looked, as Jared Diamond observes, "much as

Africa's Serengeti Plains do today, with herds of elephants and horses pursued by lions and cheetahs." Eleven thousand years ago, most of those animals were gone. Forty thousand years ago, Australia was inhabited by giant kangaroos, one-ton lizards, and four-hundred-pound flightless ostrichlike birds. Thirty thousand years ago, those animals too had disappeared.

Paul S. Martin was the first to call the extinctions of these large mammals "prehistoric overkill" and to attribute them to the hunting practices of Homo sapiens as we spread out of our African-Asian homeland into other parts of the world.* Critics of this theory argue that changes in climate might really be to blame—a period of glaciation in North America (or a quick oscillation between extreme temperatures of hot and cold), a drought in Australia, and they have found some evidence to support their position. But they haven't been able to explain why these animals were able to survive many previous periods of glaciation and drought, or why these extinctions coincided with the arrival of humans in so many different geographical locations. They haven't explained why the losses of large herbivores, herbivores weighing one hundred pounds or more, are particularly heavy in these mass extinctions. For the many reasons enumerated in this book, relating to nutrition, sex, security, and status, this is the kind of animal preferred by human hunters around the world. It is now and probably always has been.

---

*In some cases, extinctions may have been the result not of overhunting but of some indirect effect relating to human migration or population growth, such as the introduction of diseases or nonnative species like dogs or rats. Dingoes, for example, the descendants of dogs brought to Australia three thousand years ago, have had a great effect on the local marsupials. In other places, Norway rats that escaped from ships have led to scores of extinctions by outcompeting and preying on native vertebrates.

Climate may have played some role in the plight of animals during the Pleistocene Era, as it has in extinctions both before and after the advent of humans. But there is little doubt, from the large number of sites with a high density of both stone artifacts and bone refuse, that humans had developed into formidable, big-game hunters during this time and were able to take on animals that were much larger than themselves. And they had an unfair advantage in many places that probably explains why humans had so rapid and devastating an effect—and why there are still elephants and lions in Africa but not in the Americas. In parts of the world, animals had evolved, for millions of years, in the absence of humans. These animals, like the "tame" animals of the geographically isolated islands of the Galápagos—lizards that casually crawl up a visitor's leg or birds that land on a shoulder—would not have regarded the new arrivals from Africa and Asia as predators and so would have made easy prey. By the time mastodons and giant kangaroos had figured out that these new, two-legged animals should be added to their list of animals to be feared, it was too late.

More evidence for this population-driven view of history is the increases in human populations themselves. Despite the toll taken by warfare, infectious diseases, and natural disasters, our numbers have been growing, somewhat, steadily. In A.D. 1000, about the time when early inhabitants of the Georgia coast took up farming, the world population is estimated to have been 300 million. (Guesstimated really, since accurate estimates from that period are impossible.) In the early sixteenth century, when the Spanish first established colonies in the Americas, it was 500 million. By the 1900s, when millions of Chinese moved to other parts of Asia and millions of Chinese and Irish moved to the Americas, it was one billion. Today, it is six billion and is expected to reach eight to nine billion in the next fifty years.

Though significant, the changes in health that Larsen sees reflected in the bones of the people on Saint Catherines Island as they took up farming were nothing compared with the changes that accompanied the arrival of the Spanish four hundred years later.

Relatively few infectious diseases actually affect bones, so the fossils Larsen has studied from this period, found in burial sites on the grounds of Spanish missions, do not tell the tale of the terrible epidemics of smallpox and measles that devastated Indian populations in the Americas. (These diseases, brought over on Spanish ships, arose as specialized diseases of humans from germs that had first infected animals. They arose, in other words, as a result of animal domestication.) But the numerous partially severed arm and shoulder bones do document the violent nature of the early interactions between the Spanish and the Indians. These bones had felt the sharp steel of Spanish swords, for here, as in so many places in the New World, two growing populations clashed, one much more sophisticated and technologically advanced than the other.

The bones also tell what this clash between two cultures meant in terms of the daily lives of the Indians. Their diet continued to decline during the contact period, Larsen knows, because of an increase in the incidence of cavities and signs, now, of iron deficiency anemia. Many of the skulls Larsen has found from the period 1607 to 1680 are extensively pitted, a condition caused by hyperactive bone marrow and red blood cell production and indicative of an increased reliance on corn, since corn blocks the absorption of iron from most other foods. The condition of these skulls also indicates to Larsen that the Indians had reduced their intake of fish and shellfish, since these foods actually promote iron absorption when eaten with corn. Larsen doesn't know why fish were no longer as important a part of the Indian diet (a change that has been verified by chemical analyses of the bones, since marine foods leave a differ‑ ent nitrogen trace from terrestrial foods). But he suspects it had to do with the Spanish policy of *reducción*, or concentrating native populations around mission centers through forced migrations. Fishing and shellfish gathering are, after all, free‑ranging occupa‑ tions. *Reducción* had many serious health consequences. Not only did it deprive Indians of variety in their diet, but it also forced together a population that was struggling to fight off a league of newly introduced, highly infectious diseases. And it forced them to drink contaminated water from mission wells, yet another cause of iron deficiency anemia.

But *reducción* allowed the Spanish to consolidate and extend their holdings in the Americas, and that contributed to something else that Larsen is able to read from the bones: much hard labor on the part of the Indians. Historical accounts of the period note that Indians were used in construction, transportation, and growing corn for export to other Spanish missions, and that they were often subjected to harsh, abusive treatment. The fossils Larsen has found tell the same story. Signs of osteoarthritis, rare in bones from the precontact, agricultural period, now become prevalent. Over fifty percent of Indian males suffered from osteoarthritis in their backs and an even higher percentage of Indian women, though women had previously had much lower incidences of this disease than men. A consensus has emerged, Larsen observes, that it was the diseases brought by Europeans to the New World that were responsible for the large losses suffered by native Indian populations, and that "no other factor seems capable of having exterminated so many people over such a large part of North America," as one stu‑ dent of the period has remarked.

"But," says Larsen, "undernourished, overworked people are much more likely to get sick than healthy, well‑fed individuals.

Excessive workloads, reduced nutritional quality, warfare, population nucleation, social disruption, poor sanitation, and physiological stress were all things that played their part in the outcome of the interactions between European and Indian populations."

After lunch, Larsen and I walked to the lab where he stores many of his fossils, so that he could show me what diseases like tuberculosis, syphilis, and anemia—the growing pains of burgeoning populations—look like in bones. He removed a tibia, or shinbone, from a box and pointed out the deep pitting on one side. This tibia had belonged to a farmer who had lived on Saint Catherines Island hundreds of years ago and who had been suffering from an infection like syphilis, probably for decades. A fibula or splint bone from the same period had a big hole in it, which Larsen said had formed to allow pus to drain from "an especially nasty infection." A jawbone was pocked and eroded from unchecked tooth decay. As a mother, I was particularly struck by a child's skull, porous on top, with lingering signs of the infection (perhaps of meningitis) from which the child had probably died.

Humans didn't take up agriculture in order to improve their health or their life span, I reminded myself as I turned this small brown skull over in my hands. I thought about the brain that it had once held and the child who had relied upon that brain to make sense of the world, the child whose diet hadn't allowed her to fight this infection.

The good news for humans is that we've gotten better and better at producing the kinds of high-quality foods that we need to grow, thrive, and operate our big brains on. Hunger, malnutrition, and famines still exist, of course, but our capacity for producing food has greatly improved and is able to meet the world's present needs. Modern-day famines, as a Nobel Prize–winning economist recently showed, are caused more by problems with the distribution of food—that is, politics—than with food supplies themselves, a finding of small consolation to the starving individual but of great importance to policy makers.

The bad news is the tremendous toll that this inventiveness is taking on all the other animals and plants that inhabit this world. Humans are continuing to displace animals from the trees, swamps, and grasslands they need to survive. Extinctions are being reported with increasing frequency, and except for animals capable of living in our wake, animals like white-tailed deer, crows, rats, and raccoons that succeed in the kinds of habitats that zoning laws and refuse facilities create, all manner of species are being affected. As the respected Harvard biologist E. O. Wilson notes, "In the relentless search for more food, we have reduced animal life in lakes, rivers and now, increasingly, the open ocean. And everywhere we pollute the air and water and lower water tables."

Two hundred years ago, the English economist and mathematician Thomas Malthus predicted that humans would soon experience widespread famines as a result of overpopulation, and people are cheered by the failure of Malthus's dire predictions to come true. Many have come to believe that we can go on this way indefinitely, *and they gotta think what they believe.* But what Malthus didn't understand is that humans respond to population pressure with new ways of finding food. This is what humans do best and the reason, as Katharine Milton and others have suggested, for our large brains and our unusual system of food sharing. We may indeed run out of resources at some point in our future. We may eliminate so many plant and animal species that life on earth collapses. But if so, that moment is much farther off than Malthus imagined.

"Eat enough and it will make you wise," said the sixteenth-century playwright John Lyly. Smart, yes—but wise, I'm not so sure. With human populations at six billion and expected to nearly double by the year 2050, we must ask ourselves whether our large brains, which have allowed us to continually expand our resources, will also allow us to control our populations and protect the habitats of living creatures other than ourselves, to prevent our planet from being reduced to a carefully managed cafeteria for humans . . . a take-out stand for Homo sapiens alone . . . a giant McDonald's dispensing Happy Meals of near optimal amounts of proteins, fats, and carbohydrates.

Unless we value only human life, unless we can imagine wanting to live in a place so technologically complex that there is no more need for insects and bats to pollinate our fruits, no more need for plants to produce the oxygen we breathe, we must ask ourselves whether we have the wisdom to act in ways that may run contrary to our short-term reproductive goals but that consider the long-term health of the planet. Then a bird, flying over earth a thousand years from now, will see a green, inviting place with room for many of life's experiments—not just our own.

Epilogue

FORAGING IN THE GARDEN

Weeding my vegetable garden used to be a relatively simple affair. Laborious, yes, but clear cut and black and white. Out went everything that I had not grown from seed or transplanted as a tender young thing, everything that hadn't once come out of a packet. I used to take pride in my tidy rows of beans, Swiss chard, and broccoli separated by dark swaths of freshly tilled soil. An overgrown, unweeded garden felt like an unmade bed or an uncombed head of hair.

All that changed, though, when I learned to recognize certain wild plants for the delicious foods that they are. Now lamb's-quarter and purslane are allowed to flourish in and between the rows. But now, when I go down into the vegetable garden, my right hand never knows what my left hand is doing. Am I harvesting purslane's succulent leaves to make a cold summer soup, or am I pulling out the entire plant so that the basil doesn't have to compete with it for water and nutrients? Am I picking, or am I weeding? Sometimes I get so confused that I find my basket full of truly inedible weeds and my weed pile full of plants that I have grown from seed. My daughters have always been willing to help me with the garden, but now I hardly ever ask. How can I expect them to weed according to this new, evolving gardening philosophy when my own hands and mind are not up to the task?

One might ask, Are these plants worth the loss of their labor and my sense of order? Purslane is a fresh, crunchy addition to any salad. Lamb's-quarter tastes a lot like spinach but is zingier and more intense. Plus the spinach season in my part of the world lasts for about a nanosecond (in that brief, glorious moment between the rains of spring and the heat of summer). Lamb's-quarter and purslane can be harvested all summer long simply by snipping the tips off the growing stems. And these edible weeds just don't seem to feel the heat. During this summer's ruthless drought and high temperatures, when my lettuce plants lay limp and listless on the ground, they stood as tall and tasty as ever.

Healthwise, I know they're worth it. Lamb's-quarter is loaded with iron and other minerals. (So much so that I sometimes get a metallic aftertaste when I overindulge.) Purslane is one of the best plant sources of the essential fatty acid linolenic acid and is also rich in iron and vitamins A and C. By adding these plants to my family's meals, I increase the variety in our diet, so important to the health of an omnivore like Homo sapiens. I'm not promoting the consumption of all edible weeds—most are more edible than appetizing—but these two are really worth trying.

Other animals already know this, which is why I must forage for lamb's-quarter and purslane in my garden instead of in the woods or by the roads. Why I grow weeds alongside my vegetables. Deer are so numerous in my whereabouts that they quickly take care of any lamb's-quarter that dares to poke its head aboveground. Like many other flowering and nonflowering plants, it is able to grow only in the safe house, the sanctuary, of my garden and other fenced areas. Purslane is low growing so it has a better chance of escaping the attentions of deer, but still it is much more prevalent inside my garden than out.

Which brings me back to those thoughts on human population growth and overpopulation that have been dogging me ever since my visit to Clark Larsen in North Carolina. Here in New York, my neighbors can't agree on whether the white-tailed deer in our woods should be culled. So it's hard to imagine that people all over the world could possibly agree to have smaller families. It's hard to conceive of any way that humans will be able to unwind the tight spiral of human population growth: more people, requiring more resources and more technology, displacing more and more species.

These are thorny dilemmas that minds far sharper than my own have tried to address. But I do have a few thoughts after these years spent musing about food and about how the quest for food has shaped human nature. A few reasons for hope. The first has to do with the way that all animals behave under circumstances of diminishing resources. The second, with something that is unique to humans.

Fortunately, everyone does not have to agree to have smaller families in order for individuals to make this decision on their own. Natural selection favors those parents that produce the most surviving offspring. But offspring need resources in order to survive, and all animals tend to limit their broods to a number for which they may be able to acquire the necessary resources. Humans are no different. Parents, quite naturally, tend to have fewer children when the resources necessary to support children to the point of independence—clothing, food, housing, free time, college tuition, clean drinking water—become increasingly costly, that is, scarcer and scarcer.

When I was in the Arctic with Tony Gaston and his group of researchers, we spent a great deal of time talking about "breeders" and "nonbreeders" and counting the number of breeding birds in the murre colony at Coats. So we all laughed one night at dinner when I pointed out that I was the only breeder in this group. Gaston's assistants rolled their eyes at the very idea of having children, and they tried to remember the last time there had been a breeder, a human breeder, in camp—only once, several years ago. At first, it struck me as strange that these people who spend their lives studying the reproductive behavior of birds would have no interest in having children of their own. But later it made a lot of sense. If humans, like all animals, adjust the number of their offspring to their perceptions about the availability of resources, who among humans would be more aware of competition for resources than an ecologist or a biologist?

But smaller families will not by themselves be enough to solve our problems if those families consume more and more resources, as is the trend in the world's most developed nations. Humans will really be able to change the cast of the future only because of something that is uniquely human. Our human way of life may have gotten us into this situation where our numbers threaten to overwhelm all other forms of life, but it may also get us out.

For it is clear that humans *do* value life other than our own. Our sharing natures allow us to feel for other animals almost as strongly as if they were members of our own species. Our large brains allow us to understand the needs of other animals and the complex interrelationships between them, and they give us the flexibility to change our behavior. We can't rely on technology to solve our problems, since technology, as we know (like changes in diet), simply allows more mouths to be fed, but technology coupled with conservation, education, and a slower population growth may allow us to change the direction in which we are heading. We are not doomed to carry on as we have been, as members of a less intelligent, less flexible, *less sharing* species would be. Just as humans can figure out how not to overeat even when food is abundant, so we may be able to discover how not to overuse the world's resources, how to satisfy our essential appetites for food, sex, and love in a healthier way. Many thousands of years ago, men and women learned how to share food in order to survive, and the results were astonishing. Now we should learn how to share the planet.

BIBLIOGRAPHY

Aiello, Leslie C., and Peter Wheeler. "The Expensive Tissue Hypothesis." *Current Anthropology* 36:2 (1995): 199–221.

Akazawa, Takeru, and C. Melvin Aikens, eds. *Prehistoric Hunter-Gatherers in Japan.* Tokyo: University of Tokyo Press, 1986.

Altmann, Jeanne. *Baboon Mothers and Infants.* Cambridge: Harvard University Press, 1980.

Angier, Natalie. "When (and Why) Dad Has the Babies." *New York Times*, October 28, 1997, F1, F7.

Bachman, Gwendolyn C. "The Effect of Body Condition on the Trade-off Between Vigilance and Foraging in Belding's Ground Squirrels." *Animal Behaviour* 46 (1993): 233–44.

Barker, Lewis M. *The Psychobiology of Human Food Selection.* Westport, Conn.: AVI Publishing, 1982.

Barnard, Alan, and Jonathan Spencer, eds. *Encyclopedia of Social and Cultural Anthropology.* London and New York: Routledge, 1996.

Beardsworth, Alan, and Teresa Keil. *Sociology on the Menu: An Invitation to the Study of Food and Society.* London: Routledge, 1997.

Bittelsohn, Joel, et al., eds. *Listenting to Women Talk about their Health: Issues and Evidence from India.* Har Anand Publications, 1994. Distr. by South Asia Books.

Blumenschine, Robert J., and John A. Cavallo. "Scavenging and Human Evolution." *Scientific American*, Oct. 1992: 90–96.

Boesch, Christopher, and Hedwige Boesch-Achermann. "Dim Forest, Bright Chimps." *Natural History* 9 (1991): 50–56.

Boserup, E. *The Conditions of Agricultural Growth.* Chicago: Aldine, 1965.

Brody, Jane E. "Add Cumin to the Curry: A Matter of Life and Death." *New York Times*, March 3, 1998.

———. "Yes, America, You Can Get Fat on Low-fat Foods." *New York Times*, December 18, 1996.

Brown, Peter J., and Melvin Konner. "An Anthropological Perspective on Obesity." *Annals of the New York Academy of Sciences* 499 (1987): 29–46.

Byers, John A. "Play's the Thing." *Natural History* 108:6 (1999): 40–44.

Caplan, Pat. "Feasts, Fasts, Famine: Food for Thought." *Berg Occasional Papers in Anthropology 2*. Oxford: Berg, 1994.

Carpenter, F. Lynn, David C. Paton, and Mark A. Hixon. "Weight Gain and Adjustment of Feeding Territory Size in Migrant Hummingbirds." *PNAS* 80 (1983): 7259–63.

Clutton-Brock, Juliet. *The Walking Larder: Patterns of Domestication, Pastoralism and Predation*. London: Unwin Hyman, 1989.

Clutton-Brock, T. H. *Primate Ecology*. London: Academic Press, 1977.

Clutton-Brock, T. H., and Paul H. Harvey. "Mammals, Resources and Reproductive Strategies." *Nature* 273 (1978): 191–95.

Cohen, Mark Nathan. *The Food Crisis in Prehistory: Overpopulation and the Origins of Agriculture*. New Haven: Yale University Press, 1977.

Cohen, Mark Nathan, and George J. Armelagos, eds. *Paleopathology at the Origins of Agriculture*. Orlando, Fla.: Academic Press, 1984.

Crawford, M. A. "A Re-evaluation of the Nutrient Role of Animal Products." In *Proceedings of the III World Conference on Animal Production*, R. L. Reid, ed. Sydney: Sydney University Press, 1975.

Dahlberg, Frances. *Women the Gatherer*. New Haven: Yale University Press, 1983.

Damas, David. "Central Eskimo Systems of Food Sharing." *Ethnology* 11 (1972): 220–40.

Davis, Clara M. "Self Selection of Diet of Newly Weaned Infants." *American Journal of Diseases of Children* 36:4 (1928): 652–79.

De Castro, John M., and Elizabeth S. de Castro. "Spontaneous Meal Patterns of Humans: Influence of the Presence of Other People." *American Journal of Clinical Nutrition* 50 (1989): 237–47.

De Garine, Igor, and Nancy J. Pollock. *Social Aspects of Obesity*. Australia: Gordon and Breach, 1995.

Demment, Montague W., and Peter J. Van Soest. "A Nutritional Explanation for Body-size Patterns of Ruminant and Nonruminant Herbivores." *American Naturalist* 125 (1985): 641–72.

de Waal, Frans B.M. "Bonobo Sex and Society." *Scientific American* 272 (1995): 82–88.

———. "Food Sharing and Reciprocal Obligations Among Chimpanzees." *Journal of Human Evolution* 18 (1989): 433–59.

Diamond, Jared. *Guns, Germs and Steel: The Fates of Human Societies*. New York: Norton, 1997.

Dowling, John H. "Individual Ownership and the Sharing of Game in Hunting Societies." *American Anthropologist* 70 (1968): 502–7.

Draper, H. H. "The Aboriginal Eskimo Diet in Modern Perspective." *American Anthropologist* 79 (1977): 309–16.

Duffy, David Cameron. "Land of Milk and Poison." *Natural History* 7 (1990): 4–8.

Eaton, S. Boyd, Marjorie Shostak, and Melvin Konner. *The Paleolithic Prescription*. New York: Harper and Row, 1988.

Feigin, Merryl Beth, Anthony Sclafani, and Suzanne R. Sunday. "Species Differences in Polysaccharide and Sugar Taste Preferences." *Neuroscience and Biobehavioral Reviews* 2 (1987): 231–40.

Feistner, Anna T. C., and W. C. McGrew. "Foodsharing in Primates: A Critical Review." In *Perspectives in Primate Biology*, P. K. Seth and S. Seth, eds. New Delhi: Today and Tomorrow's Printers and Publishers, 1989.

Fischler, Claude. "Food Habits, Social Change and the Nature/Culture Dilemma." *Social Science Information* 19:6 (1980): 937–53.

Fisher, M.F.K. *How to Cook a Wolf*. Originally published in 1942 by World Publishing. New York: North Point Press, 1988.

———. *The Gastronomical Me*. Originally published in 1943 by Duell, Sloan and Pearce. New York: North Point Press, 1989.

Freeland, W. J., and Daniel H. Janzen. "Strategies in Herbivory by Mammals: The Role of Plant Secondary Compounds." *American Naturalist* 108 (1974): 269–89.

Galdikas, Biruté M. F., and Geza Teleki. "Variations in Subsistence Activities of Female and Male Pongids: New Perspectives on the Origins of Hominid Labor Division." *Current Anthropology* 22:3 (1981): 241–56.

Gaulin, Steven J. C., and Melvin Konner. "On the Natural Diet of Primates, Including Humans." In *Nutrition and the Brain*, vol. I, R. J. Wurtman and J. J. Wurtman, eds. New York: Raven Press, 1977.

Gill, Frank B. *Ornithology*. New York: W.H. Freeman, 1990.

Glander, Kenneth E. "The Impact of Plant Secondary Compounds on Primate Feeding Behavior." *Yearbook of Physical Anthropology* 25 (1982): 1–18.

———. "Nonhuman Primate Self-Medication." In *Eating on the Wild Side*, Nina L. Etkin, ed. Tucson: University of Arizona Press, 1994.

———. "Poison in a Monkey's Garden of Eden." *Natural History* 86 (1977): 35–42.

Good, Kenneth. "The Yanomami Keep on Trekking." *Natural History* 4 (1995): 57–64.

Goodall, Jane. "My Life among Wild Chimpanzees." *National Geographic* 124 (1963): 272–308.

———. "The Behavior of Free-living Chimpanzees in the Gombe Stream Reserve." *Animal Behaviour*. Monograph 3, 1968.

———. *The Chimpanzees of Gombe*. Cambridge: Harvard University Press, 1986.

Gordon, Kathleen D. "Evolutionary Perspectives on Human Diet." In *Nutritional Anthropology*, F. E. Johnston, ed. New York: Alan R. Liss, 1987.

Gould, Richard A. "Notes on Hunting, Butchering and the Sharing of Game among the Ngatatjara and Their Neighbors in the West Australian Desert." *Kroeber Anthropological Society* 36 (1967): 41–66.

Grady, Denise. "Nicotine's Image Takes a Turn for the Worse." *New York Times*, November 17, 1998, F1, F8.

Griffon, P. Bion, and Agnes Estioko-Griffin, eds. *The Agta of Northeastern Luzon: Recent Studies.* Cebu City, Philippines: San Carlos Publications, 1985.

Harding, Robert S. O., and Geza Teleki. *Omnivorous Primates: Gathering and Hunting in Human Evolution.* New York: Columbia University Press, 1981.

Harris, Marvin. *Good to Eat: Riddles of Food and Culture.* New York: Simon and Schuster, 1985.

Hart, John A. "From Subsistence to Market: A Case Study of the Mbuti Net Hunters." *Human Ecology* 6:3 (1978): 325–53.

Hawkes, Kristen. "Foraging Differences Between Men and Women." In *The Archaeology of Human Ancestry*, James Steele and Stephen Shennan, eds. New York: Routledge, 1996.

———. "Why Hunter-Gatherers Work: An Ancient Version of the Problem of Public Goods." *Current Anthropology* 34:4 (1993): 341–61.

Hawkes, K., J. F. O'Connell, and N. G. Blurton Jones. "Hadza Women's Time Allocation, Offspring Provisioning, and the Evolution of Long Menopausal Life Spans." *Current Anthropology* 38 (1997): 551–77.

———. "Hunting Income Patterns Among the Hadza: Big Game, Common Goods, Foraging Goals and the Evolution of the Human Diet." *Philosophical Transactions of the Royal Society London* 334 (1991): 243–51.

Headland, Thomas N. "The Wild Yam Question: How Well Could Independent Hunter-Gatherers Live in a Tropical Rain Forest Ecosystem." *Human Ecology* 15:4 (1987): 463–91.

Heinrich, Bernd, and Scott L. Collins. "Caterpillar Leaf Damage, and the Game of Hide-and-Seek with Birds." *Ecology* 64:3 (1983): 592–602.

Hill, Kim. "Hunting and Human Evolution." *Journal of Human Evolution* 11 (1982): 521–44.

Hill, Kim, and A. Magdalena Hurtado. "Hunter-Gatherers of the New World." *American Scientist* 77 (1989): 437–43.

Hill, Kim, Hillard Kaplan, Kristen Hawkes, and A. Magdalena Hurtado. "Foraging Decisions among Ache Hunter-Gatherers: New Data and Implications for Optimal Foraging Models." *Ethology and Sociobiology* 8 (1987): 1–36.

Hole, F. "A Reassessment of the Neolithic Revolution." *Paleorient* 10:2 (1984): 49–60.

Hrdy, Sarah Blaffer. *The Woman That Never Evolved.* Cambridge: Harvard University Press, 1981.

Hughes, R. N. *Diet Selection: An Interdisciplinary Approach to Foraging Behaviour.* Oxford: Blackwell Scientific Publications, 1993.

Hurtado, A. Magdalena. "My Family, Food, and Fieldwork." In *I've Been Gone Far Too Long: Field Trip Fiascoes and Expedition Disasters,* Monica Borgerhoff Mulder and Wendy Logsdon, eds. Oakland: RDR Books, 1996.

Hurtado, Ana Magdalena, Kristen Hawkes, Kim Hill, and Hillard Kaplan. "Female Subsistence Strategies Among Ache Hunter-Gatherers of Eastern Paraguay." *Human Ecology* 13:1 (1985): 1–28.

Ingold, Tim. *The Appropriation of Nature: Essays on Human Ecology and Social Relations.* Iowa City: University of Iowa Press, 1987.

Ingold, Tim, David Riches, and James Woodburn, eds. *Hunters and Gatherers. I. History, Evolution and Social Change.* New York: Berg, 1988.

Isaac, Glynn. "The Food-sharing Behavior of Protohuman Hominids." *Scientific American* 238:4 (1978): 90–108.

Jacobs, Lucia F. "Memory for Cache Locations in Merriam's Kangaroo Rats." *Animal Behavior* 43 (1992): 585–93.

Jarman, P. J. "The Social Organisation of Antelope in Relation to their Ecology." *Behaviour* 48 (1974): 215–67.

Johns, Timothy. *With Bitter Herbs They Shall Eat It: Chemical Ecology and the Origins of Human Diet and Medicine.* Tucson: University of Arizona Press, 1990.

Johnston, Francis E., ed. *Nutritional Anthropology.* New York: Alan R. Liss, 1987.

Kaplan, Hillard, and Kim Hill. "The Evolutionary Ecology of Food Acquisition." In *Evolutionary Ecology and Human Behavior,* Eric Alden Smith and Bruce Winterhalder, eds. New York: Aldine de Gruyter, 1992.

———. "Food Sharing Among Ache Foragers: Tests of Explanatory Hypotheses." *Current Anthropology* 26:2 (1985): 223–46.

———. "Hunting Ability and Reproductive Success Among Male Ache Foragers: Preliminary Results." *Current Anthropology* 26 (1985): 131–33.

Kare, Morley R., and Joseph G. Brand, eds. *Interaction of the Chemical Senses with Nutrition.* Orlando, Fla.: Academic Press, 1986.

Katz, Solomon H. "Food and Biocultural Evolution: A Model for the Investigation of Modern Nutritional Problems." *Nutritional Anthropology,* F. E. Johnston, ed. New York: Alan R. Liss, 1987.

———. "An Evolutionary Theory of Cuisine." *Human Nature* 1:3 (1990): 233–59.

Kawamura, Yojiro, and Morley R. Kare, eds. *Umami: A Basic Taste: Physiology, Biochemistry, Nutrition, Food Science.* New York: Marcel Dekker, 1987.

Kimura, Doreen. "Sex Differences in the Brain." *Scientific American,* September 1992, 119–25.

Koehler, Gary M. "Bobcat Bill of Fare." *Natural History* 12 (1988): 48–56.

Kopytoff, Verne G. "Meat Viewed as Staple of Chimp Diet and Mores." *New York Times*, June 27, 1995, C1, C10.

Krebs, J. R., and Davies, N. B. *Behavioural Ecology.* Oxford: Blackwell Scientific, 1978.

Kurland, Jeffrey A., and Stephen J. Beckerman. "Optimal Foraging and Hominid Evolution: Labor and Reciprocity." *American Anthropologist* 87 (1985): 73–93.

Larsen, Clark Spencer. *Bioarcheology: Interpreting Behavior from the Human Skeleton.* Cambridge: Cambridge University Press, 1997.

Larsen, Clark Spencer, et al. "Population Decline and Extinction in La Florida." In *Disease and Demography in the Americas,* John W. Verano and Douglas H. Ubelaker, eds. Washington, D.C.: Smithsonian Institution Press, 1992.

Laughlin, W. "Hunting: An Integrating Biobehavior System and Its Evolutionary Importance." In *Man the Hunter,* R. Lee and I. Devore, eds. Chicago: Aldine, 1968.

Leary, James R., et al. *Cross-cultural Studies of Factors Related to Differential Food Consumption.* New Haven, Conn.: HRAF, 1981.

Lee, Richard Borshay. "Eating Christmas in the Kalahari." *Natural History* 78:10 (1969): 14–22, 60–63.

———. *The !Kung San: Men, Women, and Work in a Foraging Society.* Cambridge: Cambridge University Press, 1979.

Lee, R. B., and Devore, I., eds. *Man the Hunter.* Chicago: Aldine, 1968.

Levinson, David, and Melvin Ember. *Encyclopedia of Cultural Anthropology.* New York: Henry Holt, 1996.

Lévi-Strauss, Claude. *The Raw and the Cooked.* London: Jonathan Cape, 1970.

Lima, Steven L., Thomas J. Valone, and Thomas Caraco. "Foraging-efficiency-predation-risk Trade-off in the Grey Squirrel." *Animal Behaviour* 33 (1985): 155–65.

Logue, A.W. *The Psychology of Eating and Drinking.* New York: W.H. Freeman, 1986.

Lovejoy, Q. Owen. "The Origin of Man." *Science* 211 (1981): 341–50.

Lyon, Richard H. "The Bird that Farms the Dell." *Natural History* 96 (1987): 54–60.

Marshall, Lorna. *The !Kung of Nyae Nyae.* Cambridge: Harvard University Press, 1976.

Martin, Robert D. "Human Brain Evolution in an Ecological Context." Fifty-second James Arthur Lecture, presented at the American Museum of Natural History, 1982.

McCarthy, F., and M. McArthur. "The Food Quest and the Time Factor in Aboriginal Economic Life." In *Records of the Australian-American Scientific Expedition of Arnhem Land,* ed. C. P. Mountsford. Melbourne: Melbourne University Press, 1960.

McGrew, W. C. "Evolutionary Implications of Sex Differences in Chimpanzee Predation and Tool Use." In *The Great Apes: Perspectives on Human Evolution,* vol. 5, David A. Hamburg and Elizabeth R. McCown, eds. Menlo Park, Calif.: Benjamin/Cummings, 1979.

———. *Chimpanzee Material Culture: Implications for Human Evolution.* Cambridge: Cambridge University Press, 1992.

McQuade, Denise B., Ernest H. Williams, and Howard B. Eichenbaum. "Cues Used for Localizing Food by the Gray Squirrel." *Ethology* 72 (1986): 22–30.

Mennell, Stephen, Anne Murcoll, and Anneke H. van Otterloo. *The Sociology of Food: Eating, Diet , and Culture.* London: Sage, 1992.

Milton, Katharine. "Diet and Primate Evolution." *Scientific American* 8 (1993): 86–93.

Moffett, Mark W. "Gardeners of the Ant World." *National Geographic,* July (1995): 98–111.

Murcott, Anne. *The Sociology of Food and Eating.* London: Gower, 1984.

Nishida, Toshisada. "Local Traditions and Cultural Transmission." In *Primate Societies,* B. B Smuts, D. L. Cheney, R. M. Seyfarth, and R.W. Wrangham, eds. Chicago: University of Chicago Press, 1987.

Nishida, Toshisada, et al., eds. "Meat-sharing as a Coalition Strategy by an Alpha Male Chimpanzee." In *Topics in Primatology,* vol. 1. Tokyo: University of Tokyo Press, 1992.

O'Connell, James F., Kristen Hawkes, and Nicholas Blurton Jones. "Hadza Scavenging: Implications for Plio/Pleistocene Hominid Subsistence." *Current Anthropology* 29:2 (1988): 356–63.

O'Leary, Timothy J., and David Levinson. *Encyclopedia of World Cultures.* Boston: G.K. Hall, 1991.

Owens, Mark, and Delia Owens. *Cry of the Kalahari.* Boston: Houghton Mifflin, 1984.

Pianka, Eric R. *Evolutionary Ecology.* New York: Harper and Row, 1974.

Pinker, Steven. *How the Mind Works.* New York: Norton, 1997.

Radetsky, Peter. "Gut Thinking." *Discover,* May 1995: 76–81.

Richards, Audrey. *Land, Labour and Diet in Northern Rhodesia.* Oxford: Oxford University Press, 1939.

Ridley, Matt. *The Origins of Virtue: Human Instincts and the Evolution of Cooperation.* London: Penguin, 1996.

Rodman, Peter S., and John G. H. Cant. *Adaptations for Foraging in Nonhuman Primates.* New York: Columbia University Press, 1984.

Roe, Daphne A. *A Plague of Corn: The Social History of Pellagra.* Ithaca, N.Y.: Cornell University Press, 1973.

Rolls, Barbara J. "The Role of Sensory-Specific Satiety in Food Intake and Food Selection." In *Taste, Experience and Feeding,* E. D. Capaldi and T. L. Powley, eds. Washington, D.C.: American Psychological Association, 1990.

————. "Sensory-Specific Satiety." *Nutrition Reviews* 44:3 (1986): 93–101.

Rolls, E. T., and A. W. L. de Waal. "Long-term Sensory-Specific Satiety: Evidence from an Ethiopian Refugee Camp." *Physiology and Behavior* 34 (1985): 1017–20.

Rozin, Paul, and April E. Fallon. "A Perspective on Disgust." *Psychological Review* 94 (1987): 23–41.

Schaller, George B., and Gordon R. Lowther. "The Relevance of Carnivore Behavior to the Study of Early Hominids." *Southwestern Journal of Anthropology* 25:4 (1969): 307–41.

Shipman, Pat. "Scavenging or Hunting in Early Hominids." *American Anthropologist* 88 (1986): 27–43.

Siegel, Ronald K. *Intoxication: Life in Pursuit of Artificial Paradise*. New York: Dutton, 1989.

Simoons, Frederick J. *Eat not This Flesh: Food Avoidances from Prehistory to the Present*. Madison, Wis.: University of Wisconsin Press, 1994.

Smyth, Marie. "Hedge Nutrition, Hunger, and Irish Identity." In *Through the Kitchen Window: Women Explore the Intimate Meanings of Food and Cooking*, Arlene Voski Avakian, ed. Boston: Beacon, 1997.

Speth, John D. "Early Hominid Hunting and Scavenging: The Role of Meat as an Energy Source." *Journal of Human Evolution* 18 (1989): 329–42.

————. "Early Hominid Subsistence Strategies in Seasonal Habitats." *Journal of Archeological Science* 14 (1987): 13–29.

————. "Seasonality, Resource Stress, and Food Sharing in So-called 'Egalitarian' Foraging Societies." *Journal of Anthropological Archaeology* 9 (1990): 148–88.

Speth, John D., and Katherine A. Spielmann. "Energy Source, Protein Metabolism and Hunter-Gatherer Subsistence Strategies." *Journal of Anthropological Archaeology* 2 (1983): 1–31.

Stahl, Ann Brower. "Hominid Dietary Selection Before Fire." *Current Anthropology* 25:2 (1984): 151–68.

Stanford, Craig B. "To Catch a Colobus." *Natural History* 1 (1995): 48–54.

Stefansson, V. *My Life with the Eskimo*. Originally published in 1913. New York: Collier, 1962.

Stephens, David W., and John R. Krebs. *Foraging Theory*. Princeton, N.J.: Princeton University Press, 1986.

Stevens, C. Edward, and Ian D. Hume. *Comparative Physiology of the Vertebrate Digestive System*. Cambridge: Cambridge University Press, 1995.

Stevens, T. A., and J. R. Krebs. "Retrieval of Stored Seeds by Marsh Tits Parus Palustris in the Field." *Ibis* 128 (1986): 513–25.

Strum, Shirley C. "Processes and Products of Change: Baboon Predatory Behavior at Gilgil, Kenya." In *Omnivorous Primates: Gathering and Hunting in Human Evolution*. Robert S. O. Harding and Geza Teleki, eds. New York: Columbia University Press, 1981.

Stubbs, R. J., M.C.W. van Wyk, A. M. Johnstone, and C. G. Harbron. "Breakfasts High in Protein, Fat or Carbohydrate: Effect on Within-Day Appetite and Energy Balance." *European Journal of Clinical Nutrition* 50:7 (1996): 409–17.

Taylor, Ronald L. *Butterflies in my Stomach or: Insects in Human Nutrition.* Santa Barbara, Calif.: Woodbridge Press, 1975.

Toth, Nicholas. "The First Technology." *Scientific American* 256:4 (1987): 112–21.

Trawick, Margaret. *Notes on Love in a Tamil Family.* Berkeley: University of California Press, 1990.

Trivers, Robert L. "Parental Investment and Sexual Selection." In *Sexual Selection and the Descent of Man,* B. Campbell, ed. Chicago: Aldine, 1972.

Turnbull, Colin M. *The Forest People.* New York: Simon and Schuster, 1961.

———. *The Mountain People.* New York: Simon and Schuster, 1972.

Vander Wall, S. B. *Food Hoarding in Animals.* Chicago: University of Chicago Press, 1990.

Voland, E. "Differential Parental Investment: Some Idea on the Contact Area of European Social History and Evolutionary Biology." In *Comparative Socioecology: The Behavioural Ecology of Humans and other Mammals,* V. Standen and R. A. Foley, eds. Special publication 8 of the British Ecological Society. Oxford: Blackwell Scientific, 1989, 391–403.

Whiten, A., and E. M. Widdowson, eds. *Foraging Strategies and Natural Diet of Monkeys, Apes and Humans.* Proceedings of a Royal Society Discussion Meeting held on May 30 and 31, 1991. Oxford: Clarendon Press, 1992.

Wilkie, David S., and Gilda A. Morelli. "Pitfalls of the Pygmy Hunt." *Natural History* 12 (1988): 32–40.

Williams, Florence. "A House, 10 Wives: Polygamy in Suburbia." *New York Times,* December 11, 1997, F1, F7.

Williams, G. C. "Pleiotropy, Natural Selection, and the Evolution of Senescence." *Evolution* 11 (1957): 398–411.

Winterhalder, Bruce, and Eric Alden Smith. *Hunter-Gatherer Foraging Strategies.* Chicago: University of Chicago Press, 1981.

Wrangham, Richard W. "On the Evolution of Ape Social Systems." *Social Science Information* 18:3 (1979): 335–68.

Zapalis, Charles, and Anderle R. Beck. *Food Chemistry and Nutritional Biochemistry.* New York: John Wiley, 1985.

INDEX

ABOUT THE AUTHOR

SUSAN ALLPORT is a writer specializing in history and science. Her books include *A Natural History of Parenting: A Naturalist Looks at Parenting in the Animal World and Ours*, *Sermons in Stone: The Stone Walls of New England and New York*, and *Explorers of the Black Box: The Search for the Cellular Basis of Memory*.

Allport has contributed essays, travel articles, and other pieces to the *New York Times*, the *International Herald Tribune*, the *Hartford Courant*, *Audubon* magazine, *Connoisseur*, the *Providence Journal-Bulletin*, and the *Missouri Review* and she lectures at the American Museum of Natural History and other locations. She lives with her husband and two daughters in Katonah, New York.